DAVID JOHNSTON | TOM JENKINS

How
Canadian Innovators
Made the World

n!ous

Smarter, Smaller, Kinder,
Safer, Healthier, Wealthier,
and Happier

SIGNAL
McCLELLAND
& STEWART

Hardcover edition published 2017

Signal and colophon are registered trademarks of
McClelland & Stewart

Library and Archives Canada Cataloguing in Publication
is available upon request.

Library of Congress Control Number is available upon request

ISBN: 978-0-7710-5091-6

Printed and bound in Canada

Published by Signal,
an imprint of McClelland & Stewart,
a division of Penguin Random House Canada Limited,
a Penguin Random House Company

www.penguinrandomhouse.ca

1 2 3 4 5   21 20 19 18 17

In honour of the relentless creativity of all of the people who have inhabited this mighty and graceful land we call Canada.

CONTENTS

INNOVATION: HERE'S HOW
Proven approaches to innovation that anyone can use

SMARTER

SMALLER

HEALTHIER

WEALTHIER

HAPPIER

Why we wrote this book.

This book is about innovation, not invention. While invention is often an element of innovation, it is never the whole story. In our experience, the whole story is usually more compelling.

We think the information in this book is going to come as a big surprise to many people. Perhaps you're among them. If we had asked you yesterday to name Canadian innovations now in worldwide use, would your mind have leapt instantly to the electric light bulb, the propeller, the electric radio, the United Nations Declaration of Human Rights, recycling, movie theatres, nuclear physics, *and* Superman? Pretty soon you'll never forget them and the many other innovations in this collection that show Canadians at their most creative.

Ours is an era of uncommon opportunity. New ways of meeting, problem solving, designing, and delivering what people need and want have blown the doors off our old ways of thinking and collaborating. The technologies of computing and communications have now intersected at every corner of human endeavour. There has never been a better time to have an idea, share it, improve it, and turn it into something that changes something else for the better. Action or reaction, product or process, habit or habitat, everything is ready for change.

Canada has a long tradition of welcoming new ideas and, at this stage in our national development, we urgently need better ideas in every arena. We want to help. So we wrote this book to celebrate the history and spirit of creativity in Canada, and to inspire all Canadians to think of themselves as innovators with a critical role to play in the improvement of everything we do together as a society, both at home and throughout the world. The first step in that process is to allow yourself to be inspired. And there is nothing that will do the job faster than reading about people just like you who have harnessed their curiosity and creativity to improve what they saw around them with astounding impact. And they are Canadians.

We gathered these stories with the help of our friends and colleagues in industry, government, community organizations, universities and colleges, national institutions, and of course from within our own circles of family and friends. The collection is necessarily small and certainly personal. Although it's not exhaustive, we have worked hard to make sure it illustrates the range and depth of innovations that Canadians have given to the world.

A notable feature of innovation is simultaneity. People in different parts of the world routinely get the same idea at approximately the same time. As a result, many people can legitimately claim bragging rights for having had that idea first. We expect that our book will lead to lively debate about who did what and where and exactly when. When those conversations occur, innovation will become an everyday, commonplace topic. That is our fondest hope.

This volume is part of a larger national effort to build the first-ever collection of Canadian innovations. As we go to press, there are already more than one thousand other innovations ready to be enjoyed at www.innovationculture.ca, and we hope that with your help that number will grow quickly. Log on. Enjoy. Tell the world. It's time everyone knew that Canadians make the world a smarter, smaller, kinder, safer, healthier, wealthier, and happier place.

David Johnston & Tom Jenkins
Ottawa & Waterloo, Canada
March 2017

What makes Canadians so innovative?

E very human is creative. Every community is inventive. The history of our species is the long story of innovation itself. Great ideas spring up everywhere, often in remote corners of the planet, firing our imaginations and making the impossible seem possible at last. We are all capable of hatching fresh ideas for solutions to age-old problems. Yet some nations have populations that seem to be particularly good at framing problems in ways that make creative thinking easier. Canada is one of those nations. Why?

As we move from example to example in this small and illustrative collection of innovations, a few answers rise almost naturally off the page. The first is rooted deep in the history of the land. For thousands of years, the Indigenous inhabitants of the northern half of North America explored, cultivated, and shared the land we now call Canada. Their experience, like ours today, was of a vast collection of widely different topographies and climates. To cope, they innovated systems of language, community, hunting, agriculture, transportation,

manufacture, defence, art, commerce, and spirituality. So refined were these by the time Europeans arrived that such iconic innovations as canoes (*p. 40*) and kayaks (*p. 42*), moccasins (*p. 176*) and snowshoes (*p. 43*), toboggans (*p. 43*) and dogsleds (*p. 42*) were adopted by the new-comers, either immediately or after practical experience proved the merit of these innovations. The story of the incursion into Aboriginal lands over the past five hundred years has many shameful, lamentable chapters, but it can be said albeit with sad irony that the success of newcomers in great measure depended on the adoption of the vast set of innovations engineered by generations of First Nations communities long before any immigrant man or woman set foot on these shores. As we relearn our history as Canadians, we will uncover and make possible more innovation, more invention, drawing on the Indigenous civilizations about which we still know so little. Just as Canada is a work in progress, so too is this kind of compendium about our ingenuity. Our story can only get better. In our day, there are some positive indications of a return of respect for the wisdom of our Indigenous hosts in this land, and the adoption of more recent Aboriginal innovations stands among them. The use of justice that is restorative (*p. 79*) rather than punitive was directly inspired by First Nations practices and has been used in Canadian justice systems now for over forty years. New sustainable communities (*p. 86*) designed to address critical housing inadequacies in the north are now based on

legacy Inuit knowledge of changing climate and respect for the unique traditions of community that evolved in their land. New expressions of frustration, anger, sorrow, and loss related to the disappearance of thousands of Aboriginal women across Canada in recent decades now find innovative expression in memorials (*p. 89*) that honour a growing Indigenous assertion of identity, spirituality, activism, and loss.

Climate has been another defining factor. While some of the country enjoys warm and even hot summers, Canadian winters are an event unto themselves. It surprises many visitors to the nation's capital city that the temperature in Ottawa has been recorded as high as +40°C in summer and as low as -40°C in winter, the greatest fluctuation of any world capital. Samuel de Champlain and his band suffered so badly in the Bay of Fundy during the winter of 1604 that they abandoned their original island colony and moved to a location out of the wind just to survive (*p. 176*). Many of the first settlers to Halifax froze to death their first winter because during the warm summer they simply couldn't believe how cold things might get. *Habitants* in Quebec learned to construct roofs at sharply oblique angles to let the annual masses of snow slide to the ground rather than weigh down their houses. At some points, the weather has been cold enough to make growing crops impossible right through the summer. Heavy snowfalls were reported in Nova Scotia, New Brunswick, Quebec, and Ontario in June of 1816, with frost and light snow continuing into July. No wonder that local farmers have been eager to devise hardier, faster-growing crops such as Marquis Wheat (*p. 148*) and the McIntosh apple (*p. 177*). Weatherstripping to prevent icy drafts (*p. 197*) and thermal windows (*p. 188*) to insulate without blocking the view were both Canadian responses. Little surprise too that Canada's contribution to seasonal innovation includes buffalo coats (*p. 174*) and long johns (*p. 72*), the snowmobile (*p. 58*) in all its forms, the snowblower (*p. 102*), and the system of words that the international scientific community now uses for the many different kinds of snow they encounter (*p. 23*).

Between 1600 and 1900, a good proportion of settlers to Canada were trained mariners who brought people and supplies to our shores, and who thereafter kept the sea lanes of commerce and communication open. Seafarers by necessity are an innovative sort. A ship at sea is a world unto itself, and sailors must become adept at using what is on hand and on deck to repair something broken or improve something that is not serving a need. When those people began inhabiting the small towns of our Atlantic provinces, they turned in self-reliance to the needs ashore — transportation, agriculture, construction, communication, and the like.

Among their contributions were the propeller (p. 44), foghorn (p. 94), dinner theatre (p. 176), and kerosene (p. 95). Readers of this volume who notice the rich variety of innovations from the East Coast will have good cause to remember the great strides made possible because of Canada's maritime heritage. Farther west, the cultivation of the Prairies by small populations looking out at long horizons of arable land required new machines that could make farming on this new scale possible. Students of agricultural innovation will note that the self-propelled combine harvester (p. 158), roto thresh combine harvester (p. 127), air seeder (p. 165), rod weeder (p. 156), and rotary car dumper (p. 20) were innovations that transformed western Canada from a community of self-sufficient farms into a bread basket for the whole world.

Coming from the crowded cities and stratified social structures of Europe, early settlers found themselves in dispersed small towns, depending on and yet physically distant from their neighbours, especially in winter. This provided a twin impulse toward innovation. First, they had to devise their own solutions to local problems. In these isolated environments, any practical idea was acceptable. Second, when short of expertise, they had to seek it either within easy means of travel or through a rich exchange of correspondence in which ideas were shared, considered, and improved. By the late 1800s, this desire for intellectual connection led to an innovation of its own – the research team. In stark opposition to the life of the lone inventor, Canadian innovators around the time of Confederation knew well that they had to collaborate to innovate. One of the first and most successful experiments in creating a hub of skills was Alexander Graham Bell's team of like-minded tinkerers who gathered at his family estate at Beinn Bhreagh near Baddeck in Cape Breton, Nova Scotia. While wildly creative himself, Bell was humble. (He often admitted his ignorance of electrical engineering, claiming that his invention of the telephone [p. 48] sprang rather from his knowledge of speech pathology.) So he built a team to refine his ideas and find better solutions by using their own specialized knowledge.

This early notion of reliance on team was distinctly Canadian. Copied around the world, it has evolved in our own time into a concept now referred to as the hub. The hub is a point of connection among teams with complementary skills and resources willing to collaborate on a common objective. The point of connection can be geographic, as it is in Waterloo, Ontario, whose businesses, universities, and associations have enjoyed extraordinary success in taking a good idea, making it great, finding a product or service that makes best use of it, then mustering the finances to manufacture, package, brand, market, and support that idea as it moves forward. The two-way pager (p. 67), the search engine (p. 206), and the BlackBerry (p. 31) all came from the same place because it was possible to incubate a great idea in that hub.

Another uniquely Canadian context for innovation is an inherent national tendency to want to get along. Canadians, who smilingly acknowledge the fact, are routinely ribbed for their civil obedience, politeness, and, sorry to say, a tendency to over-apologize. British comedian Jimmy Carr recently quipped that it is unsatisfying playing to Canadian audiences because his act depends on interaction with hecklers and Canadians don't heckle; they just smile as he makes fun of them. We know that behind that politeness is a root belief that harmonious relationships are critical; Canadian history has proven that teamwork inevitably produces better results than ego. Our great leaps in avalanche protection (*p. 106*), wheat production (*p. 148*), telesurgery (*p. 143*), and space exploration (*p. 28*) to name just four have all depended on open-minded, long-term collaboration among large numbers of individuals and organizations, many of whom have never been and perhaps never will be publicly celebrated for their contributions.

INVENTION OR INNOVATION?

In Canada's centennial year of 1967, a meticulous narrative history of advances in technology appeared called *Ideas in Exile: The History of Canadian Invention*. The title hints at author J.J. Brown's central thesis that while Canadians were ingenious, they were so in isolation. The subtitle teases out the root of the problem; their

pursuit was invention not innovation. The difference between the two concepts is small but critical. Invention (from the Latin *invenire* meaning *coming in* or *arriving at*) is really the act of finding out, of discovering, whether accidentally or as a result of search and effort. The concept of invention is exciting to us because it describes the bold, individual search for truth that, over the years, led to the Age of Enlightenment in the 18th century and subsequently gave birth to all the modern sciences. This span of time was concurrent with the evolution of Canada into a nation of its own. Throughout, we've been fascinated by the successes of all those lone geniuses whose innate curiosity led them to reject established beliefs and build new understanding based on systematic observation, measurement and experiment, and the formulation of hypotheses that in turn could be tested and refined. These techniques, which we now casually lump together under the term *scientific method*, required the creation of whole new categories of instruments, machines, processes, and approaches that we still refer to as *inventions*, simply because of the way they were arrived at. A number of these have been Canadian and are celebrated on the following pages, including the roller bearing (*p. 12*), steam buggy (*p. 47*), radio voice transmission (*p. 52*), walkie-talkie (*p. 58*), gas mask (*p. 101*), G-suit (*p. 125*), and pacemaker (*p. 129*). We all enjoy reading about inventions, but our long-standing focus on ideas conceived in exile by solitary geniuses has perhaps led us to presume that the great advances are made only in exile and only by geniuses. That is not the case.

That's why the term *innovation* is so helpful. To innovate (from *innovare*, meaning *renew* or *alter*) implies a deliberate change in the nature or fashion of something, precisely to make it more useful to more people. To the innovator, impact is the ultimate measure of success. Innovation has always been far more common and, until lately, far less written about than pure invention. Anyone can innovate. We're all curious. We're all creative. When we share our ideas and refine them together, we all have the power of those lone geniuses. When we refuse to act in isolation, when we move away from the presumption that great ideas are conceived in exile, together we become ingenious, which we think is a much better title for a book in our day.

Innovation is the creative combination of anything that, once done, makes something better. So Robert Foulis (*p. 94*) did not *invent* the foghorn that dreary night in 1853 as he walked back to his home in Saint John. He simply heard the sound of his daughter playing the piano as he approached the house and picked up the low notes first. He wasn't the first human to have observed that low-frequency sounds travel farther than high-frequency sounds, but given his background he was able to put that fact into a new perspective. Foulis lived in a community of mariners who all risked running aground on foggy nights when lighthouses could not be seen. So he put two and two together. If ships could be warned by sound as well as light, fog wouldn't be so perilous. And if the warning could be made with sound, it would ideally be both loud and at low frequency to reach ships when they were as far away from the danger as possible. How to make a sound loud? Steam! Again, Foulis had not invented the steam whistle, but he knew that steam produced louder sounds than all other technologies. But wait. Even then, if sailors heard a faraway blast, how would they know which point of land it had come from? Foulis hadn't invented Morse code either (that innovation had already been around for at least a decade) but he knew that combinations of dots and dashes could communicate volumes of complex information. So he refined his idea, thinking that a coded, low-frequency steam whistle would likely make seafaring safer. And it did. That novel combination of simple ideas in a chain produced the foghorn, which is still in use all around the world with countless lives saved every year because of it. Invention? Hardly. Brilliant? Certainly. Of course, to have impact, the foghorn project needed a vast collaboration of engineers, builders, seaside property owners, cartographers, factories, governments, and, of course, shipping companies and mariners who worked hard to put it into practical use. Only then did it prove its obvious benefit and inspire others to adopt the innovation worldwide.

Jump ahead more than a century to 1985. In her Toronto home, Wendy Murphy (*p. 80*) was greatly upset as she watched the news on television. A massive earthquake had devastated Mexico City. The quake was of such severity that the death toll eventually climbed to more than ten thousand. Wendy watched as rescue teams pulled survivors from the rubble and observed how many infants and young children had to be carried to safety one by one. Her question that

night was the first question most innovators ask: *Could this not be done a better way?* Wendy's simple idea was an apparatus by which rescuers could move bundled babies to safety at the same time. It was in essence a long stretcher on which rescue teams could strap up to six infants for rapid evacuation. The folding apparatus, evocatively called WEEVAC, was novel yet uncomplicated. But to take that idea from a single creative spark into a commonly used solution required collaboration with teams of designers, material experts, manufacturers, marketers, distributors, and agencies of first responders around the world. In Wendy's case, it also inspired the creation of a company to make and sell the product, ensuring that the WEEVAC improved stretcher would have its intended, life-saving impact.

This theme of iterative improvement rises off every page of this book. Benjamin Tibbets (*p. 11*) did not invent the steam engine, but in 1853 he did something about that machine's colossal waste of energy by building a second cylinder that could be driven by the steam exhausted but still hot from the first. It was his compound steam engine that went on to revolutionize the transportation industry. Sheila Watt-Cloutier (*p. 140*) didn't invent climate study, she wasn't the first to link climate change to disease, and she certainly wasn't the only observer to ascribe climate change to industrial emissions. She was, however, the first to combine all three in a legal petition against the United States of America, claiming that the Inuit right to a clean, stable climate had been denied by that government's neglect. Nyle Ludolph (*p. 80*) didn't invent the processes to recycle glass, plastic, and paper. He didn't pioneer door-to-door pickup. And he didn't know himself what colours of plastic were most resistant to sunlight. But as a garbage collector he knew that his town's landfill site was maxed out, and he figured that communities right across Canada were facing the same issue, so he simply suggested that some garbage could be collected from people's houses and recycled instead of thrown out with the regular trash. Blue-box recycling, now a worldwide phenomenon, was the direct result. And while Elsie MacGill (*p. 23*) was indeed the world's first woman to earn degrees in electrical and aeronautical engineering, and the world's first woman to design an aircraft, she was not the first person to use mass-production techniques to build vehicles with rigorous quality control. Elsie's innovation was to borrow those techniques from the automotive industry she knew well and apply them to the aviation industry in 1943. It is that innovation that quickly became the norm worldwide.

On the pages that follow, you will learn about hundreds of Canadian innovations, all of which have given useful benefits to people facing particular problems. Even though space is tight, we've tried to show something about the initial spark that launched the innovation, often just a small event, single insight, or happy accident. We've tried to reveal how teamwork is often part of the original thinking and *always* part of the ultimate solution. And we've organized the book to categorize these innovations by the impact they've made in the world. Thanks to Canada, the world is smarter, smaller, kinder, safer, healthier, wealthier, and happier than before.

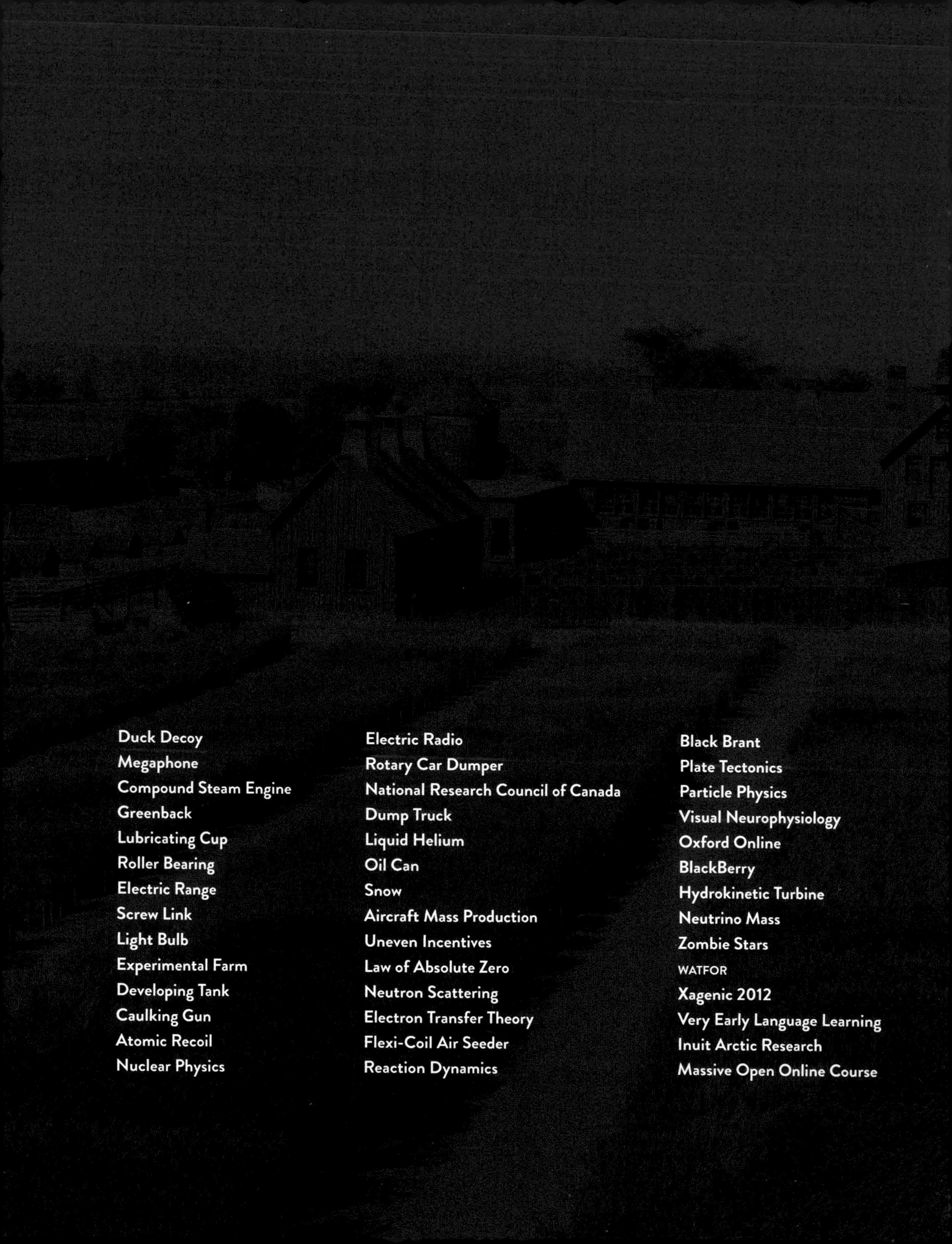

Duck Decoy
Megaphone
Compound Steam Engine
Greenback
Lubricating Cup
Roller Bearing
Electric Range
Screw Link
Light Bulb
Experimental Farm
Developing Tank
Caulking Gun
Atomic Recoil
Nuclear Physics

Electric Radio
Rotary Car Dumper
National Research Council of Canada
Dump Truck
Liquid Helium
Oil Can
Snow
Aircraft Mass Production
Uneven Incentives
Law of Absolute Zero
Neutron Scattering
Electron Transfer Theory
Flexi-Coil Air Seeder
Reaction Dynamics

Black Brant
Plate Tectonics
Particle Physics
Visual Neurophysiology
Oxford Online
BlackBerry
Hydrokinetic Turbine
Neutrino Mass
Zombie Stars
WATFOR
Xagenic 2012
Very Early Language Learning
Inuit Arctic Research
Massive Open Online Course

ONE

Smarter

Canadians find it fitting that the most common visual metaphor for a truly innovative idea is an illuminated light bulb. After all, the electric light bulb itself is a Canadian innovation. From ideas as simple as the dump truck to those as profound as nuclear physics, teams of Canadian innovators have made it possible to move dirt, produce energy, reduce friction, develop photographs, plant seeds, listen to radio, look up words, lubricate machinery, debug code, construct aircraft, observe the earth, understand the stars, and even talk about the weather in ways that are smarter than anything that came before.

Smaller
Kinder
Safer
Healthier
Wealthier
Happier

DUCK DECOY
The hunter's secret weapon.

The hunter's most formidable weapon is deception. The Cree and Ojibway peoples of Canada's Great Lakes relied on it for thousands of years. They used reeds, cattails, bulrushes, tamarack, and other plants to make remarkably lifelike floating and stationary decoys that lured game birds and waterfowl to roosting areas. Once there, they were within reach of the nets, snares, arrows, and spears of the Aboriginal hunters. European settlers and then generations of recreational hunters wisely took up the practice for themselves – a deceptively simple technique that continues to this day in much the same way it has for thousands of years.

MEGAPHONE
The best way to call a moose.

Why change something that's worked perfectly for thousands of years? Today's moose hunters have no reason to adapt the megaphone used by their Ojibway and Attiamek predecessors. Made out of birchbark, bound with spruce roots, and secured with leather straps, these devices amplify and direct the sound of the moose call, attracting the creatures to the hunters. While today's versions may be made out of different materials – plastic and whatnot – the enduring Aboriginal megaphone is still the best way to call a moose.

COMPOUND STEAM ENGINE
The more efficient generator.

Innovation is not usually invention. The compound steam engine is a perfect example of how smart thinking can make a good thing even better. In this case, the thinker was Fredericton, New Brunswick, native Benjamin Tibbets. The problem he considered was the wastage of heated steam in engines. Before Benjamin put his mind to the problem, steam engines required vast amounts of carbon fuel to produce vapour, which was then used briefly to produce power and subsequently exhausted into the atmosphere. In 1853, he built a new kind of steam engine incorporating a reservoir and a second cylinder. These two features enabled the engine to take the steam discharged from the main high-pressure cylinder and put it to further use as low-pressure steam. The result was a more efficient power generator that used much less fuel per unit of usable energy. The concept, refined by others, was subsequently incorporated into all steam engines with immediate impact. The additional power made developments such as long-distance train travel a reality at last.

GREENBACK
The un-counterfeitable dollar.

Why are American dollar bills green? The answer came from McGill University in Montreal in 1861. That year, the United States Congress authorized production of $50 million in demand notes, largely to cover the federal government's burgeoning war expenses. The next year, Thomas Sterry Hunt – an American chemist working for the Canadian Geological Survey at McGill – came up with the idea of using chromium sesquioxide for banknote ink. Hunt discovered that, when used as ink for the notes, the compound's green pigment could not be bleached by acid, thus making it incredibly durable. Nor could it be photographed, making it impervious to counterfeiting. Hunt's innovation was the straightforward use of science to solve an everyday problem. And with his discovery, the un-counterfeitable dollar – the greenback – was born. In Canada.

LUBRICATING CUP
The real McCoy.

It was so much better than all the other kinds that railroad engineers asked for it by name. The "it" is Elijah McCoy's automatic lubricating cup for oiling the steam engines of locomotives and ships. When the mechanical engineer developed it in 1872, McCoy's cup was a boon to railroaders throughout the world. The device supplied lubricating oil to the cylinders, bearings, and axle box mountings of locomotives automatically. This method boosted productivity, enabling trains to run faster and more profitably, because they didn't need to be stopped en route to their destinations to have their engines greased. Over the years, McCoy refined his signature device and created new ones, ending up with fifty patents related to his lubricating devices. In a pleasing symmetry, McCoy was born of another kind of railroad. He was the son of slaves who escaped to Canada via the Underground Railroad and settled in Colchester, Ontario, near Windsor. A son of one railroad, a father of another. He's the real McCoy.

ROLLER BEARING
The friction reducer.

Canada's greatest contribution to the design of modern machines is the roller bearing. Created by George Thomas of Digby County, Nova Scotia, in 1879, the bearing reduces if not eliminates the friction generated by the contact of moving machine parts with stationary ones. The device does so by housing rollers or ball bearings within a roller cage. Today, roller bearings are more important than ever. They are small yet critical components of a variety of devices — from bicycles and cars to farm equipment and industrial machinery. These friction reducers mean smooth running for the world's machines.

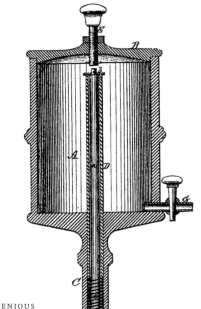

ELECTRIC RANGE
The power cooker.

Thomas Ahearn's dinner guests were a contented lot that Ottawa evening of 1882. They had just finished an elaborate meal prepared for them by their fellow electrical engineer. Their mood changed from satisfied to horrified when their host revealed he had cooked the feast using electricity. Developed in secret, Ahearn's electric range used resistance coils to convert electricity into heat. A full ten years after this inaugural, magical demonstration, the first commercial electric oven was installed in Ottawa's Windsor Hotel. It was slow to catch on elsewhere. Towns and cities had to be electrified before homes could be equipped with the new stoves, so it wasn't till the 1930s that electric ranges began to replace their gas cousins. After that, Ahearn's innovation quickly made the wood stove obsolete. And Canadians proudly recall that the age of electric cooking was born right in Ahearn's own Ottawa kitchen.

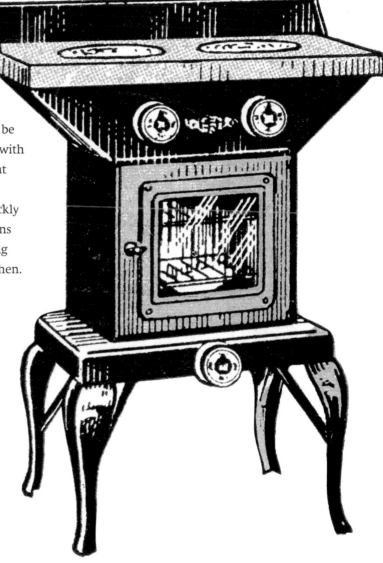

SCREW LINK
The mighty bond.

Take a stroll through your local hardware store and you'll find a simple fastener that's as useful and ubiquitous today as the day it was created. Developed by Donald Munro of Pictou, Nova Scotia, in 1885, the screw link is that simple, enduring fastener. It's just an open, rigid link in the shape of the letter C with an exterior thread on one end and an interior thread on the other. A sleeve with the opposite threads closes and opens the link as needed. Turn the sleeve in one direction to expose the opening; turn it the other to close. The screw link made rural life much easier, enabling farmers to hitch and unhitch heavy equipment in seconds without worrying that something might slip. Simple and secure, the mighty bond is as strong as ever.

LIGHT BULB
The bright future.

Thomas Edison didn't invent the electric light bulb. Credit for that illuminating discovery must go to an unlikely duo from Toronto. Dreaming of a bright future in 1874, medical student Henry Woodward and hotelkeeper Mathew Evans fashioned a bulb out of a glass tube that contained a large piece of carbon connected to two wires. When they hit the switch, the current flowed and the carbon glowed. But not for long. They then filled their bulb with inert nitrogen to prolong the burn. After testing the bulb with repeated success, the Torontonians patented

their invention in both Canada and the United States and then confidently set out to secure the financing needed to take their bright idea from lab to marketplace. As has so often happened in Canada, they failed. Time and time again they were told their idea was just too risky. Unfunded, they looked for a buyer of both the U.S. patent rights for their bulb and the exclusive licence to their Canadian patent as well. The entrepreneurial Edison was quick to close the deal. The Wizard of Menlo Park then refined their design and unveiled his own miraculous invention in a dramatic display on New Year's Eve 1879. The subsequent electrification of the world made this Canadian innovation one of the most important contributions to modern life. In all the jubilation, only two men were left alone in the dark.

EXPERIMENTAL FARM
The centre of agricultural science.

For many generations and especially in its earliest years, Canada was a nation of farmers, its economic success propelled by their work — especially in the west. The Central Experimental Farm, located in Ottawa, was established in 1886 to carry out crop research to support them, and to serve as the focal point for a nationwide system of farms devoted to experimental research. The new farm made its presence felt immediately by developing Marquis wheat, a variety that matures early to suit western Canada's shorter growing season. Researchers at the farm have worked steadily since to use agricultural science to shape and improve farm life — and therefore all life — in Canada. The farm itself has also evolved. Originally situated on the outskirts of the city, it now sits in the centre, an oasis of rolling fields, ornamental gardens, and arboretum — a garden devoted to trees. It is a hub for both research and recreation, its fields, paths, and hills playgrounds for skiers, cyclists, and tobogganers. Canada may no longer be primarily a farming nation, but this advance in agricultural science remains a vibrant centre.

DEVELOPING TANK
The portable darkroom.

If we can pinpoint the moment photography exploded as a hobby, the year is 1899. Working with a hand-picked team, Arthur McCurdy of Baddeck, Nova Scotia, created the photo-developing tank that year. The invention by Alexander Graham Bell's one-time private secretary made it possible for the developing process of film to be carried out in daylight. The enclosed tank not only protected the film from damaging light, but also prevented the chemical contents within from spilling out when agitated. In so doing, the tank made photography less the exclusive domain of professionals and more a pastime that could be enjoyed by anyone with a yen to capture the world around them. Many more famous innovations came out of research at Baddeck in the Bell years, but remember this one: it's when ordinary people began to share their vision of the world in a whole new way.

CAULKING GUN
The high-speed healer.

Theodore Witte didn't merely find fresh bread and tasty desserts at his local bakery. The Chilliwack, British Columbia, inventor gained inspiration. Watching a baker wield a cake-decorating gun to ornament a sweet treat, Theodore took the idea and applied it to develop what he called a puttying tool. His 1894 creation is what we know today as the caulking gun. Just place a tube of caulking putty in the gun's ratcheted piston. Now squeeze the trigger-like handle and watch as the piston forces the putty out through a nozzle attached to the end of the tube. When traced along the edge of a window, Theodore's puttying tool applies a weatherproof seal accurately and evenly. Just make sure to keep your puttying tool out of the kitchen.

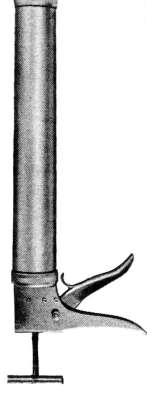

ATOMIC RECOIL
The truth about radioactive decay.

You've likely never heard of Harriet Brooks, which is unfortunate. The first Canadian woman to become a nuclear physicist, she laid the groundwork for nuclear science. In 1901, while working as a post-graduate student under the guidance of Ernest Rutherford at Montreal's McGill University, she conducted a series of experiments to determine the nature of radioactive emissions from the chemical element thorium. Her pioneering research enabled her to discover the chemical element radon, a rare radioactive gas, and to determine its atomic mass. She also uncovered atomic recoil. Also known as atomic decay, recoil is a kind of radioactive decay in which an atomic nucleus emits an alpha particle and thereby transforms into an atom with a mass number that is reduced by four and an atomic number that is reduced by two. Despite making these penetrating observations into nuclear radioactivity and transmutations, her career was short-lived. In 1907, she married and left physics, since it was mandatory in universities back then for any woman to resign from her job when she married. Plain old sexism: that's likely why you've never heard of Harriet Brooks.

NUCLEAR PHYSICS
The logic of the atom.

Ernest Rutherford is widely considered the father of nuclear physics. His defining accomplishment came in 1917 when he split the atom in a nuclear reaction. In doing so, he discovered and named the proton, and blasted the world into the atomic age. This breakthrough didn't happen in a vacuum. Rutherford carried out much of his early work in nuclear physics at Montreal's McGill University from 1898 to 1905. While there, he conceived the idea of radioactive half-life, proved that radioactivity involved the nuclear transformation of one chemical element to another, identified alpha and beta radiation, and outlined differences between the two. These achievements earned him the Nobel Prize for chemistry in 1908. More importantly, they set the stage for even more earth-shattering atomic discoveries to come.

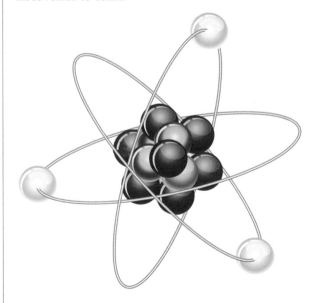

ELECTRIC RADIO
The freedom from batteries.

Today the name Rogers is synonymous with communication in Canada. While most associate the surname with the TV and mobile empire, the first man behind the name was a radio guy. And not just any radio guy. In 1925, Edward Rogers Sr. developed the first commercially viable all-electric radio in Toronto, Ontario. The tireless inventor also created an adaptor set that made it possible for owners of old sets to throw away their batteries and plug their radios into the nearest light socket. Radio batteries weren't like today's small, powerful varieties. Before young Ted's all-electric set came along, radios were powered by bulky rechargeable batteries that stained rugs and floors with leaking acid. Their annoying background hum also marred reception and sound quality. The Rogers Experimental Tube 15S — a vacuum tube for radio — enhanced the sound quality of radio, made the fledgling medium a dependable one, and shaped the future of communications not just in Canada but around the world. And within only ten years, the leaky radio battery was nothing but an unpleasant memory.

ROTARY CAR DUMPER
The railway's salute to gravity.

A powerful cabinet minister for more than twenty years, Clarence Decatur Howe is known primarily as Canada's Minister of Everything. Did you know he was also an inventor? In 1914, young C.D. was living in Fort William, Ontario (now part of Thunder Bay), where he worked as chief engineer of the Board of Grain Commissioners, the organization responsible for the regulation of the grain-handling industry in Canada. Along with supervising the building of terminal elevators, the enterprising engineer designed the first rotary car dumper. This special grain-handling car relied on gravity rather than brute force to empty its contents. It did so in eight minutes. Before the advent of Howe's car, the same operation took a crew of twenty men an hour to complete. As the whole country would one day come to learn, the Minister of Everything wasn't a man to waste time.

NATIONAL RESEARCH COUNCIL OF CANADA
The temple of science.

If innovation is triggered by women and men working across disciplinary borders and administrative walls, then a national organization to foster collaboration across walls and borders is paramount. Call it an innovation organization. Canada's is the National Research Council (NRC). One of the first of its kind in the world, the NRC was created in 1916 to promote and coordinate research to support Canada's war effort. Located in Ottawa, the organization quickly became the country's prime gathering place for experts from a variety of fields to solve really hard problems, reveal fresh thinking, and expand the frontiers of human understanding in subjects from the structure of the atom to the ruling principles of the cosmos. So impressed was a foreign visitor by what he saw that he dubbed the NRC a temple of science. More than a hundred years after its birth, it remains a temple of science, an innovation organization devoted to bridging people and disciplines, and revealing the wonder of discovery. More important, the Council has been a non-stop generator of great ideas, marketable technologies, and healing therapies, all meticulously set out in volumes of successful patents and celebrated with countless awards and honours.

DUMP TRUCK
The quick spill.

Perhaps the greatest time-saver for the modern labourer is — of all things — the good ol' dump truck. Think about it. Instead of needing a group of strong backs to shovel a big load of dirt or gravel or whatever out of the box of a truck, the dump truck just, well, dumps it. Credit for the first one goes to Robert Mawhinney. In 1920, the Saint John New Brunswicker put together a truck equipped with a special dump box in back. The dump box was fitted with a mast, cable, and winch. A simple crank handle was used to operate the winch, which tugged on the cable that lifted the front end of the box high enough to dump its load out the open back. His idea was an instant hit; within a decade the dump truck was mandatory equipment wherever earth was moved. Shovels down, lads.

GET READY TO INNOVATE: HERE'S HOW.

☐ Admit that everyone is creative.

☐ Ask people about their own innovative ideas, and let their thoughts inspire you.

☐ Keep a notebook of innovative ideas in all areas that interest you.

☐ Talk to people about your ideas and be guided by their feedback.

☐ Build a team, knowing that good ideas become great when people collaborate.

☐ Expect great results. Your belief in the possible is the sure route to innovation.

☐ Be persistent, knowing that all successful innovations have had many prototypes.

LIQUID HELIUM
The secret of superconductors.

Sometimes the genesis of innovation is just plain old luck. Gordon Shrum was a veteran of the Battle of Vimy Ridge whose schooling had been interrupted by the World War One. Back home in Toronto and strolling by the University of Toronto physics department one day in 1920, he considered how he might be able to return to his studies and earn a degree. There on the roadway he bumped into a friendly sort who turned out to be John McLennan, the department head. When he heard Gordon's story of action at Vimy, McLennan offered the man a job on the spot. Not sweeping floors. McLennan wanted Shrum to do nothing less than liquefy helium. Turns out the job needed someone who wasn't afraid of being in the middle of an explosion or two. While McLennan's team had a process in mind to transform the element from gas to liquid, violent eruptions had occurred during their experiments. They needed someone who wouldn't flinch. Three years and the odd explosion later, the job was done. The discovery enabled researchers to study the properties of metals at extremely low temperatures, which has led most notably to the advent of superconducting magnets used in processes such as magnetic resonance imaging. Lucky Gordon, lucky us.

OIL CAN
The crude dropper.

Ernie Symons was a tinkerer. The Rocanville, Saskatchewan, handyman repaired busted machines, made all sorts of tools out of pieces of scrap, and always looked for ways to take something – anything – and change it so it worked better. When a neighbour gave him three pump oilers, Ernie did what came naturally – he got to work to improve them. The main feature of his new oil can pump was a long spout that allowed it to oil machines easily, especially those hard-to-reach spots. It worked so well he filed for and was granted a patent in 1922. Then he opened up shop, manufacturing more than 3,500 during the next few years. Business really took off during the Second World War. Some 65,000 of Ernie's oilers were sold in 1943 alone. Canadian servicemen from Saskatchewan reported seeing them at work in Italy, Germany, and even Burma. When the war ended, Symons Metalworkers kept humming, making not only the company's standard oil cans but also specialized ones for various industries for the next six decades.

SNOW
The vocabulary that built a science.

Snow is just snow, isn't it? Think again: the physical and mechanical properties of snow vary widely, and this divergence has implications on how people travel through it and live and operate in it. The first to understand this thoroughly were Canada's Inuit, who developed a vast vocabulary to distinguish one snow type from another. George Klein leaned on this Inuit knowledge and added to it the rigour of modern science. Alongside his many other innovations, this prolific Canadian mechanical engineer led the effort to create the world's first instruments and protocols to measure and describe the hardness, depth, and surface qualities of snow. Klein presented his breakthrough in snow science in 1948 at the Oslo Snow Conference of the International Union of Geodesy and Geophysics National. Three years later, the International Classification for Snow was officially sanctioned. Still in use today, the classification gives us a measurable, standardized way to improve the design and safety of all kinds of infrastructure — from bridges to buildings to energy, transportation, and communications systems. It's a vocabulary that began with the Inuit of Canada's North and built a science to make our world safer.

AIRCRAFT MASS PRODUCTION
The queen of the hurricanes.

Elsie MacGill's life was one of firsts: the first woman in Canada to earn a degree in electrical engineering; first woman in the world to be awarded a master's degree in aeronautical engineering; first woman to design a plane; first woman to hold the position of chief of aeronautical engineering at an aircraft company; and the brains behind the world's first mass-produced aircraft. Soon after World War Two erupted in Europe, Elsie's Canadian Car and Foundry in what is now Thunder Bay, Ontario, was selected to manufacture the Hawker Hurricane for the Royal Air Force in 1939. Elsie swung into action. She took control of production, streamlined operations to churn out increasing numbers of aircraft, and even designed a series of modifications to equip the fighter for cold-weather flying. By 1943, the company had produced more than 1,400 Hurricanes and its workforce had grown from 500 to some 4,500 — more than half of them women. Elsie had devised and perfected the mass production of aircraft, a mode of production that was soon the norm worldwide. No wonder this Canadian woman of firsts was crowned "Queen of the Hurricanes."

UNEVEN INCENTIVES
The value of knowing more or less.

When William Vickrey was asked why he had
received the Nobel Prize, he couldn't come up
with an answer. The citation stated that the
award was for "his fundamental contributions to
the economic theory of incentives under asymmetric information." Born in Victoria, British
Columbia, the unassuming economist countered
that all he tried to do in his professional life was
carry economic theory to its logical conclusions.
Dr. Vickrey began his career in 1946 and spent
the next fifty years doing just that – using
abstract economic theory to uncover solutions to
everyday problems. Those solutions cover a
range of subjects as varied as transit fares and
traffic tolls, taxation systems, and the price of
public utilities. Wonder why you pay less money
when you use power during non-peak hours?
That's William Vickrey's work. Wonder why you
pay a lower fare for the subway or train outside
of rush hour? That's William Vickrey's work.
Wonder why cities around the world are studying a toll system to place a social cost on com-
muting and thereby ease congestion in city
centres? That's William Vickrey's work. It's
brilliant new thinking that improved the
workday lives of millions of people – from a man
who admitted that he couldn't explain what his
Nobel citation meant.

LAW OF ABSOLUTE ZERO
The new understanding of thermodynamics.

The third law of thermodynamics states that the
entropy (the unavailability of a system's thermal
energy for conversion into mechanical work) of
a perfect crystal at absolute zero is exactly equal
to zero. William Giauque set out to prove that
this law is a basic natural law. His investigations
led to two breakthroughs in 1926. The Niagara
Falls, Ontario, native discovered oxygen isotopes
17 and 18 in the earth's atmosphere – a finding
that revealed that physicists and chemists had
unwittingly been using different scales of atomic
weight. He also determined the entropies and
thermodynamic properties of many condensed
gases. And he developed the process of cooling
by adiabatic demagnetization. This magnetic
refrigeration got him closer to absolute zero
than anyone before. The achievement enabled
scientists to gain a greater understanding of the
principles and mechanisms of electrical and
thermal conductivity, determine heat capacities,
and investigate the behaviour of superconductors
at extremely low temperatures. Dr. Giauque's
pioneering work into the third law of thermodynamics earned him the Nobel Prize in chemistry
in 1949. It's the highest prize for a man who
went lower than ever before.

NEUTRON SCATTERING
The crystal-clear solution.

Often the best way to uncover how something works is to smash open that something and see what gets scattered about. This principle applies to even the smallest somethings in the universe—atoms. To do the delicate bashing, physicist Bertram Brockhouse created the triple-axis spectrometer. It measures the energy of neutrons after they have been beamed through and scattered by a crystal. The beaming through causes the neutrons to collide with and deflect off the crystal's atoms. Dr. Brockhouse called it neutron scattering. This deflection sets off vibrations, and the frequency of the vibrations can be used to calculate the forces between the atoms within that substance. Dr. Brockhouse,

who carried out much of his pioneering work in neutron scattering from 1950 to 1962 at Atomic Energy of Canada's Chalk River Nuclear Laboratory, used the triple-axis spectrometer to reveal vital properties about the structure and behaviour of atoms. This knowledge and the tool he created are essential to our understanding of solid-state physics and organic chemistry. Neutron scattering is still used today by biologists to study the elemental structure of viruses and DNA molecules. Dr. Brockhouse was granted the 1994 Nobel Prize in physics for developing neutron scattering and spectroscopy. That's something really big for understanding something really small.

FREE YOUR CREATIVITY: HERE'S HOW.

☐ Be active in disciplines other than your own.

☐ Go for a walk, knowing you'll get your best ideas away from the computer.

☐ Doodle. Make a mess and let your mind follow the pen or pencil.

☐ Write ideas within circles connected by lines to other ideas that relate.

☐ Roll large marbles or pebbles in your left hand to stimulate your creative right brain.

☐ Write on paper first, and refine your wording with a word processor only later.

ELECTRON TRANSFER THEORY
The urge to jump.

Electron transfer is the simplest and most elemental form of a chemical reaction. It makes breathing possible. It is the very essence of plant growth. Without it, life is simply impossible. The process consists of one outer-sphere electron transferring between substances of the same atomic structure. In 1951, Rudolph Marcus of Montreal was the first person in the world to record mathematically how the overall energy in a system of interacting molecules changes to induce an electron to jump from one molecule to another. His finding was a huge leap forward in the theoretical knowledge of chemical reactions. It also propelled researchers across all branches of chemistry to uncover dramatic new insights into corrosion, photosynthesis, electrical conductivity in polymers, and many other complex chemical reactions. Anyone who has used any modern implement made of metal or plastic has benefited from Marcus's work, as these materials are now stronger, more flexible, and more durable than before. For his achievement in furthering the understanding of the athletics of electrons, Dr. Marcus was awarded not Olympic gold but the Nobel Prize.

FLEXI-COIL AIR SEEDER
The gentle seed feed.

Inspiration struck Emerson Summach while he watched his young son play in the family garden. The little lad was dragging an old coil spring through the soil, leaving a diagonal pattern in it. Emerson got to thinking: Could I apply what my boy was doing to seed my own fields? The Asquith, Saskatchewan, farmer took his idea for a coil design packer to the big equipment manufacturers. When they turned him down, telling him they had no machine to make the large coils he had in mind, he made his own machine. By 1952, Emerson and his brother Kenneth had set up shop and in five years' time had built and sold four thousand of his flexi-coil air seeders. The sudden popularity of the new machine was due to its many benefits. It could be pulled behind conventional equipment without needing a set of wheels. It had no spokes to get clogged with stones and moist soil. Most of all, it didn't pulverize the land. Instead, it followed the contours of the field, leaving a diagonal pattern in the soil that slowed erosion during the spring runoff or heavy rains. More than sixty years later, this gentle seeder is a fixture in farmers' fields around the world.

REACTION DYNAMICS
The brilliant insight.

The light first went on in 1952. While working at his lab in Ottawa, John Polanyi used spectroscopy (the investigation and measurement of light when matter interacts with or emits electromagnetic radiation) to examine vibration and rotation in iodine molecules. The light grew brighter in 1958. While working at the University of Toronto, the chemist made his discovery of chemiluminescence, the light emitted by a molecule when it is in an excited state. He did this while observing the exothermic reaction of molecular chlorine with atomic hydrogen. The light grew even brighter over the years to come. So much so that by 1986 Dr. Polanyi had developed the technique of infrared chemiluminescence. The method measures weak infrared emissions from newly formed molecules in order to examine the energy given off during chemical reactions. For shedding a brighter light on the dynamics of chemical reactions at the molecular level, he received the Nobel Prize for chemistry. Like Dr. Polanyi himself, the committee recognized brilliance when they saw it.

BLACK BRANT
The rocket with atmosphere.

Within two years of the launch of Sputnik, Canadian engineers had blasted off a sounding rocket that aimed much higher. Its name was Black Brant. Launched in 1959, the missile was developed in Valcartier, Quebec, by a large consortium whose members included the Canadian Armament and Development Establishment, the Defence Research Board, and Bristol Aerospace. The research rocket carried a payload of instruments used to carry out scientific experiments and observations of the earth's upper atmosphere. Results of the experiments and readings from the observations were then tele-metered back home to earth. Even the rocket's distinctive colour had a purpose: Brant was black so researchers could monitor the rocket's roll and pitch while in flight. Used repeatedly by Canadian and American space agencies since that first flight, more than eight hundred Black Brant rockets of various kinds were launched, making them the most popular and successful sounding rockets ever built. Nice try, Sputnik.

PLATE TECTONICS
The earth-shaking idea.

Believing the earth's crust is one big rocky piece is the equivalent of believing the earth is flat. And yet it wasn't until relatively recently that we came to understand that the rigid outer layer of the earth is broken up into many moving pieces. The concept is known as plate tectonics. Its originator is John Tuzo Wilson. After working for years with a multidisciplinary team at the University of Toronto, the Canadian geologist and geophysicist first theorized in 1962 that the crusty lithosphere (one of two outer layers of the earth) is made of separate tectonic plates that ride on the weaker, fluid-like asthenosphere (the other outer layer). Wilson's revolutionary insight gave us a whole new understanding of the very nature of our planet, and of volcanoes, earthquakes, and continental drift. It has also led us to appreciate that seven plates make up the bulk of the continents and the Pacific Ocean. His idea now may seem as obvious as saying the earth is round, but it is a truly earth-shaking insight. The scientific understanding that followed has allowed a comprehensive reinterpretation of the causes of earthquakes, volcanoes, and tsunamis and has saved tens of thousands of lives through better prediction of major seismic upheavals.

PARTICLE PHYSICS
The trail of the quark.

Just when we thought matter didn't get any smaller. Everything is made up of atoms, which have a nucleus of protons and neutrons surrounded by electrons. In the early 1960s, Canadian physicist Richard Taylor and a team of trusted colleagues at Stanford University in California began a series of experiments to delve even more deeply into the makeup of atoms. Their work involved taking high-energy electrons and crashing them into protons and neutrons. Sounds like fun. The results revealed that not only did the electrons scatter, but also other particles were produced. But what exactly were these particles? By 1967, Dr. Taylor and

company were ready to find out. Taking advantage of a powerful new accelerator, the team smashed protons and neutrons to pieces. They found that these elementary particles were indeed made up of *quarks*, a reality that had been proposed three years earlier by physicist Murray Gell-Mann. Their discovery of proof of these particles within the particles is now considered a critical aspect of what is known as the standard model of matter. The Nobel Prize committee agreed. In granting Drs. Taylor, Friedman, and Kendall its 1990 award for physics, the committee called their find a "new rung in the ladder of creation."

VISUAL NEUROPHYSIOLOGY
The attack on cataracts.

Astounding results occur when experts from different countries work together to tackle shared problems. Our current understanding of how we see can be attributed to the work of a two-person Swedish-Canadian research team. The Canadian half is David Hubel of Montreal, the Swedish half Torsten Wiesel originally of Uppsala. Their twenty-year research partnership, begun at Harvard University, brought about the first major breakthroughs in visual neurophysiology. In 1978, their pioneering experiments uncovered how the brain uses a variety of detectors — edge, motion, colour, and stereoscopic depth — to process signals from the eye. Their work made it possible for us to understand and treat cataracts and strabismus in children. The duo is rightly recognized as one of the most dedicated and insightful research teams in visual neurophysiology and, as their Nobel prizes attest, in all of science. Proof that innovation is borderless.

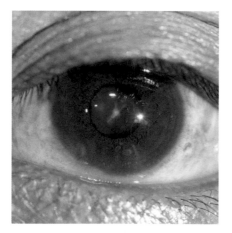

OXFORD ONLINE
The meaning of everything.

A digitized volume of more than 600,000 words and their meanings that can be searched in less than one second: that's the definition of the Oxford English Dictionary Online. At the heart of the OED online is its search engine. Created in 1989 by computer scientists Frank Tompa, Gaston Gonnet, and Tim Bray at the University of Waterloo, this mechanism made an essential resource instantly and wholly accessible. It also spawned a company as OpenText further developed the OED online search engine to be one of the world's first search engines for the web. The OpenText Web Index could find, search, and retrieve the contents of every document on the web, not just keywords and leading phrases as before. OpenText is one of several companies that together have made Waterloo an innovation hub. Experts from many disciplines converge here to share and refine ideas and find meanings to the most perplexing questions anywhere. Want to know more? Type Waterloo into your search engine, and thank Frank, Gaston, Tim, and OpenText that you can.

BLACKBERRY
The catapult to the future.

Nineteen ninety-six stands as the opening year in our era of anytime-anywhere communications. That year, Mike Lazaridis unveiled the prototype of his first wireless communications device. Sure, there were all kinds of cellular phones on the market by then, but while most telecom firms and mobile carriers still focused on voice, the enterprising engineer from Waterloo, Ontario, believed people were just as eager to communicate by data. The next year, his company, Research In Motion, released the BlackBerry, the first mobile device synched with its owner's email account. Yet it wasn't until 2002 that Lazaridis's creation truly exploded on the scene. Sleek and colourful, this latest version of the BlackBerry device enabled its owners to communicate via email and phone. It also incorporated BlackBerry Messenger, a proprietary service by which people could confidently exchange text messages over a dedicated network in complete privacy. The device was a communications marvel and, within months, a cultural phenomenon. Slick and sophisticated, the BlackBerry was soon in the palms of seemingly everyone, at once fuelling and satisfying their blossoming addiction to being connected anytime, anywhere. The device that took shape in the mind of Mike Lazaridis will long be remembered as the Canadian innovation that launched the age of mobile digital communication.

HYDROKINETIC TURBINE
The friendly power.

Rivers never stop running. This truth makes flowing water the most reliable source of power. More than wind, more than sun, more than tides. In 2005, Clayton Bear, a Calgary entrepreneur, used this knowledge as the inspiration to develop his hydrokinetic turbine. Put simply, the device is a vertical turbine that rotates and thereby generates power as water flows through it. Suspended between two mooring points, the portable water turbine can be placed wherever a particular river runs fastest. It doesn't need to be anchored to a river bottom, where flow may not be so strong. Nor is it a device that requires massive supporting infrastructure that distorts or damages the natural environment. Equally important, the compact turbine is lightweight and easy to assemble, which makes it ideal for use in remote and hard-to-reach regions, especially those in developing countries. Now communities in even the most remote parts of the world will have access to affordable power. It may just be the new energy our planet has been waiting for.

NEUTRINO MASS
The underground stargazer.

One of the most important discoveries about the nature of elementary particles was made in 2001 two kilometres underground at the Sudbury Neutrino Observatory (SNO Lab, pictured here) a detector facility located down a mine just outside the northern Ontario town. Arthur McDonald of Queen's University in Kingston, Ontario, and his team of physicists determined that electron neutrinos from the sun were oscillating into muon and tau neutrinos. Neutrinos, which come in electron, muon, and tau varieties, are elementary particles created in nuclear reactions such as those that take place in the sun. In uncovering neutrino oscillations, the Canadian team revealed that neutrinos have mass, something that had been thought impossible. Their finding also explained the lower than expected numbers of neutrinos observed from the sun. For years, theoretical models had predicted that the sun should be making neutrinos in staggering numbers. Yet detectors had seen far fewer than expected, leading some physicists to speculate that those so-called missing neutrinos had changed, or oscillated, into muon or tau neutrinos and had thus far eluded detection. Dr. McDonald transformed speculation into verification, enhancing our understanding of the nature of the sun, modifying the standard model of elementary physics, and earning himself the 2015 Nobel Prize in physics.

ZOMBIE STARS
The night of the dead.

She is the world's foremost authority on zombies. Stars, that is. Victoria Kaspi of Montreal's McGill University studies the extreme physics of neutron stars. Known as zombie stars, each one forms when a massive star runs out of fuel and explodes in a supernova but hasn't yet collapsed to the point of becoming a black hole. Why study zombie stars — other than because they have a strange name? Because they have equally bizarre properties that make it possible for scientists to test physical theories that are impossible to try out on earth. Zombie stars, for instance, are so dense that a teaspoon of one weighs a billion tonnes. That's billion with a b. In 1999, Dr. Kaspi used her laboratory in the heavens to discover the fastest-spinning pulsar — one that rotates at 716 times per second. Pulsars are thought to be rapidly rotating neuron stars. She found only the second magnetar in our galaxy. A magnetar is a neuron star that has a colossal magnetic field. And she used binary pulsars — two neutron stars orbiting each other — to test Einstein's general theory of relativity. Dr. Kaspi proves there is life in the night of the dead. Her work is not just theoretical; insights gleaned from her studies of zombie stars are now being used to help find habitable planets other than our own. No wonder Victoria Kaspi has been made a Fellow of the Royal Society and been awarded both the Prix Marie-Victorin of the province of Quebec and the Gerhard Herzberg Canada Gold Medal for Science and Engineering, the first woman ever to be so honoured.

WATFOR
The clever compiler.

It was a team victory. In the summer of 1965, four undergraduate students at the University of Waterloo were frustrated by the complexity of programming in FORTRAN, at the time the monarch of programming languages. What bugged them most was the slow and unreliable system of error diagnostics. Working with Professor J. Wesley Graham and Peter Shantz, the four students – Gus German, James Mitchell, Richard Shirley, and Robert Zarnke – built a better FORTRAN compiler for the IBM 7040 computer called WATFOR. What for? Like all innovation, just to make something better. Their compiler was faster and more accurate at error diagnostics than FORTRAN at both critical stages of programming – compiling and execution. Brilliantly marketed by their colleague Sandra Bruce, the new compiler became a hit and was soon used by programmers in more than seventy-five institutions around the world.

```
*****DEBUG***** UNIVERSITY OF WATERLOO *****
            $JOB    WATFIV
       1            ISUM=C
       2            X=1
       3      5     ISUM=ISUM + X
       4            X=X+1
       5            IF(X.LE.1000)GOTO5
       6            PRINT.ISUM
       7            STOP
       8            END

            $ENTRY
        5CC5CC

CORE USAGE          OBJECT CODE =      328 BYTES.

DIAGNOSTICS         NUMBER OF ERRORS=          0

COMPILE TIME=       0.03 SEC.EXECUTION TIME=
```

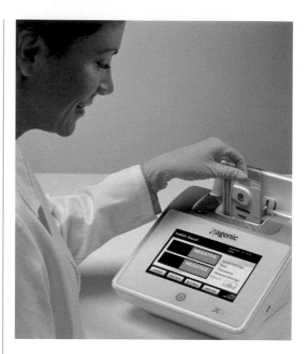

XAGENIC 2012
The instant diagnoser.

Imagine being able to diagnose cancer or some other deadly disease almost immediately. You don't need to imagine; the method has arrived. It's called Xagenic X1. Developed in 2012, the diagnostic device takes advantage of a nucleic acid detection technology to perform molecular assays of patients. A nucleic acid is a complex organic substance present in living cells whose molecules consist of many nucleotides linked in a long chain. An assay is a procedure to measure the biochemical or immunological activity in a molecular sample. Created by Xagenic, a Toronto company led by Dr. Shana Kelley, the product gives doctors their patients' diagnoses within twenty minutes, enabling physicians to make treatment decisions for their patients right away, improving care, reducing costs, and even saving lives. Imagine that.

VERY EARLY LANGUAGE LEARNING
The infant science.

Talking to infants long before they can understand their first word is an essential part of acquiring a language, and learning two languages simultaneously from birth is as natural as learning one. These two conclusions are the work of Janet Werker. In coming to them, the director of the University of British Columbia's Infant Studies Centre has pioneered a new field of study: very early language learning. Honoured for her work in 2015 with the Social Sciences and Humanities Research Council of Canada Gold Medal, Dr. Werker modestly credits her success to the sophisticated tools now available to researchers, which not only enable her to

arrive at groundbreaking answers, but also empower her to pose ever more profound and penetrating questions. Some of these have touched on the effect a mother's depression has on the ability of her child to develop language and the beneficial effects that treatment for depression can have. To make her observations, Dr. Werker focused on how children actually watch and listen to their mothers. This close study has produced a number of innovative insights. The ambition she holds for her work is equally deep and thoughtful: to understand what it means to be human and use that knowledge to advance the human condition.

INUIT ARCTIC RESEARCH
The seen and the unseen.

The best scientific research combines what can be seen and what can't. Inuit fishermen in Canada's North work closely with university researchers from the country's south to carry out this kind of research to gain a deeper understanding of the Arctic. Joey Angnatok is one of them. A fisherman from Nain in Labrador, Angnatok is the owner of MV *What's Happening*, a coastal and offshore marine-research vessel that observes and records the effects of sea-ice changes on local populations of flora and fauna in the vast expanse from Goose Bay, Newfoundland and Labrador, to Resolute,

Nunavut. Angnatok and his crew not only collect data for studies into sea ice, estuaries and fjords, and persistent organic pollutants and metals in fish and seals, but they also supply traditional and first-hand knowledge to scientists about local land, waterways, animals, and environmental changes. This blend of old and new, seen and unseen is producing new findings that reveal the often hidden effects of pollution and climate change on the Arctic, and the repercussions of these effects on the most visible feature of the region – the people who live there.

MASSIVE OPEN ONLINE COURSE
The big class.

A MOOC sounds as if it's an imaginary creature spawned from the mind of Lewis Carroll or J.R.R. Tolkien. Yes, it's big and powerful. But no, it's not lurking in a swamp or under your bed. A MOOC is a massive open online course. Massive because it can grow to be any size. Open because virtually anyone can take part in it. Online because it takes place across many web-based communications platforms and services. Course because it fulfils curriculum requirements or, better yet, enables teachers and students to share and learn and arrive at a greater understanding together. In all these ways, MOOCs supercharge practical yet played-out ideas such as distance learning, correspondence courses, and broadcast lectures. Dave Cormier is the man behind the menacing acronym. The University of Prince Edward Island educator and researcher coined the term in 2008 to label a course that was a literal expression of massive, open, and online: the course dealt with connectivism and connective knowledge; although it was offered by the University of Manitoba, it was led by instructors from two other institutions – George Siemens of Athabasca University and Stephen Downes of Canada's National Research Council. Participants included 25 tuition-paying students and some 2,200 people who took part online and paid nothing. Today, countless MOOCs exist, lurking not so menacingly within educational organizations, companies, and networks throughout the world.

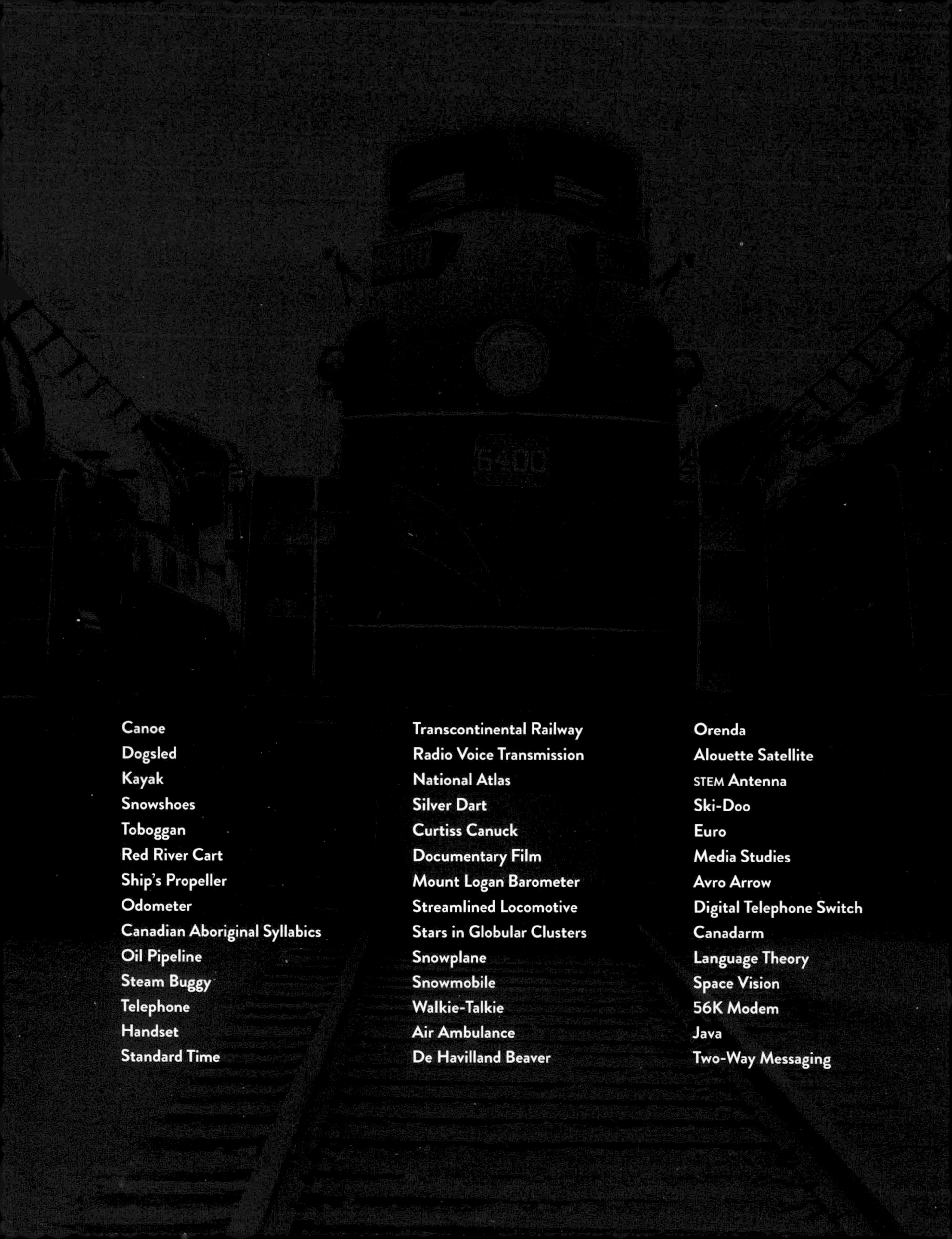

Canoe	Transcontinental Railway	Orenda
Dogsled	Radio Voice Transmission	Alouette Satellite
Kayak	National Atlas	STEM Antenna
Snowshoes	Silver Dart	Ski-Doo
Toboggan	Curtiss Canuck	Euro
Red River Cart	Documentary Film	Media Studies
Ship's Propeller	Mount Logan Barometer	Avro Arrow
Odometer	Streamlined Locomotive	Digital Telephone Switch
Canadian Aboriginal Syllabics	Stars in Globular Clusters	Canadarm
Oil Pipeline	Snowplane	Language Theory
Steam Buggy	Snowmobile	Space Vision
Telephone	Walkie-Talkie	56K Modem
Handset	Air Ambulance	Java
Standard Time	De Havilland Beaver	Two-Way Messaging

Little surprise that so many ingenious ways to close the distance between people were first devised in Canada—the world's second-largest country. Transportation has been a Canadian obsession for thousands of years, since the time when canoes and kayaks skimmed the watery highways that linked our communities. The ship's propeller, steam buggy, snowmobile, railroad, and air ambulance join countless examples of vehicles that helped bring people together and make the world smaller. In the modern era, distance has been redefined with another Canadian specialty: communications. The telephone, walkie-talkie, talking radio, two-way pager, digital telephone switch, and BlackBerry were all pioneered in Canada and then quickly adapted wherever people wanted to close the gap.

Smarter
Smaller
Kinder
Safer
Healthier
Wealthier
Happier

CANOE
The all-Canadian transport.

Is there any form of transport more Canadian than a canoe? The country's Aboriginal peoples evolved this unique form of an ancient watercraft as the ideal vehicle for a territory riddled with lakes and rivers. Light, fast, quiet, easily built from materials at hand, customizable for a thousand uses, the Canadian canoe (as it is still known by the International Canoe Federation) is a marvel of lightweight simplicity and power. When European explorers and adventurers first fetched up on Canadian shores, they learned the art of canoe making from their new-world hosts and then relied on these rapid craft as they mapped the very outline of the country to come. With his First Nation guides, Samuel de Champlain canoed all the way to Georgian Bay in 1616. Alexander Mackenzie, the first European to cross North America, did so by canoe in 1793.

Meriwether Lewis and William Clark followed in 1804, as did David Thompson in 1811, paddlers all. For two hundred years, accounts of the ingenious innovation excited European respect for the engineering prowess of Canada's first peoples, a respect that continues to this day.

RICE LAKE CANOES.

								Price.
No. 1—28 in. wide, 13 ft. long, weight about 40 lb., rib and batten							....	$22
" 2—28 "	15	"	"	"	55	"	"	24
" 3—31 "	16	"	"	"	75	"	"	26
" 4—32½ "	17	"	"	"	85	"	"	28

								Price.
No. 1—28 in. wide, 13 ft. long, weight about 40 lb., Herald's Patent							....	$32
" 2—28 "	15	"	"	"	55	"	"	34
" 3—31 "	16	"	"	"	75	"	"	36
" 4—32½ "	17	"	"	"	85	"	"	38

The above prices represent the Canoes built with iron nails, finished with iron, and painted.

When ordered to be built with copper nails, finished with copper, and varnished, an extra charge is made of $7 each on the Rib and Batten Canoe, and $10 each on the Herald Patent.

If requested, each Canoe could be made from one to two inches deeper and wider, on the same length for the same price.

D. HERALD, BUILDER,

Gore's Landing P. O.   Rice Lake, Ont., Canada.

DOGSLED
The un-improvable transport.

Invented by the Inuit thousands of years ago, the dogsled seems an unlikely candidate for transportation flawlessness — no hybrid engine, no antilock brakes, no five-speed transmission. So thought the Royal Canadian Mounted Police. In 1948, the Mounties were set to expand operations into the Far North and, knowing they would be relying on dogsleds, asked Canada's National Research Council to conduct tests on how to improve the design and construction of this ancient conveyance. The council could not. Researchers put the conveyance through the testing wringer and found it to be near flawless. The only drawback: the sled's wooden bottom could catch on high, sticky snow, jerking the sled to a stop. But the possible modification — a bottom panel made out of Bakelite — was worse than the potential ailment. If the Bakelite bottom broke on a frozen northern stretch, where was an officer going to find a replacement? The conclusion was that through centuries of small refinements, the Inuit had devised the perfect transportation for their environment. The Inuit had known that all along.

KAYAK
The high-speed hunter.

The kayak is *the hunter's boat* in name; it means exactly that in Inuktitut, the language of its creators. The kayak is also the hunter's boat in design; it is fast and manoeuvrable, used by Inuit hunters with equal effectiveness on rivers, inland lakes, and coastal waters. The kayak is old. Inuit hunters have relied on them for at least four thousand years. The classic vessel is constructed entirely out of natural materials, made of stitched sealskin or the skins of other animals stretched over a frame made of wood or whalebone. The cockpit is covered with a jacket of skin that creates a waterproof seal and secures the paddler if the kayak capsizes in rough waters. Each kayak is unique to its owner. While the concept of the kayak may be uniform — a small, narrow boat with a cockpit in which a hunter sits and that is propelled by means of that man equipped with a double-bladed paddle — each kayak is fitted to the dimensions of the person who propels it. Traditionally made by males, each kayak was three times as long as its builder's arms stretched out, as wide at the cockpit as his hips plus two fists, and as deep as his fist plus an outstretched thumb. Today, the kayak is used far and wide by women and men for many purposes. Kayaks are made to race in whitewater, surf big waves, navigate sea swells, and just glide along. The high-speed hunter has also been used by British commandos and U.S. Navy Seals in covert operations, often dropped by parachute into operation with its paddlers.

SNOWSHOES
The winter walkers.

Look to the snowshoe hare. Indigenous peoples of Canada no doubt did when they designed their first snowshoes. Just as this species uses the large surface area of its oversized feet to traverse quickly through deep snow, so did Canada's Indigenous peoples rely on their bigger-than-usual shoes to hunt. That doesn't mean all snowshoes were created equal. Sizes and styles differed according to local conditions: Inuit of the Far North relied on a circular shoe to handle deep powder; Cree of the Prairies developed a longer and narrower version for the windswept plains; while Iroquois of the Canadian Shield depended on much shorter ones to manoeuvre in crowded forests. European settlers to Canada got into the act on their arrival. Snowshoes were essential equipment for traders, adventurers, trappers, and lumberjacks. Snowshoeing soon became a fashionable winter pastime—first in Montreal and then throughout the rest of the country. Today, snowshoeing remains popular, with styles and designs of winter walkers increasingly sophisticated and specialized.

TOBOGGAN
The first snowmobile.

For thousands of years, the First Nations peoples of Canada's Far North required few possessions to live. All they needed to survive could be pulled—by person or dogsled—from place to place on a toboggan. Seven to ten feet long and narrow enough to slide within sled and snowshoe trails, toboggans are made using two or more thin boards of larch or birch wood attached by crossbars. The boards are turned up at the front—the curling achieved by bending the wood while it is wet or green and then lashing it into position until dry. When they arrived in the north, European traders, hunters, and trappers adopted the toboggan as their own. Settlers in various parts of the country soon found they could also use toboggans for fun. They started tobogganing clubs, built elaborate slides, and even inspired new sports—luge, bobsled, and skeleton—which were important enough to feature in the Olympics. Today, people throughout Canada's North still use toboggans, often in tandem with snowmobiles: the modern snowmobile and the first snowmobile working together.

RED RIVER CART
The long-distance freighter.

How do you hold together the parts of a wagon or cart when iron nails are either nowhere to be found or far too expensive to use? The Métis people of what is now the Canadian West built their Red River carts using no iron at all — only wood and animal hides. Inspired in design by the two-wheeled charettes used in New France, Red River carts were built as of 1801 using local materials only. Wooden mortices and tenons held together the wooden rails, rear boards, floorboards, and crosspieces of the carts. A mortice is a hole or recess cut into the wood. A tenon is a corresponding projection of another piece that fits into the mortice so as to join or lock the parts together. Wooden axles were lashed to the carts using strips of bison hide, also known as shaganappi, which were attached when wet. The strips then shrunk and tightened as they dried. This all-natural design made the carts buoyant. They could be floated across rivers and streams. Yet they were also rugged and strong. Pulled by horses or oxen, Red River carts could haul loads as heavy as half a tonne. Even when the carts broke down, their all-natural materials made them easy to repair. All you had to do was remember to pack a supply of wood and shaganappi — the Red River equivalent of a spare tire and jack.

SHIP'S PROPELLER
The end of sail.

Perhaps the most consequential event in modern nautical history occurred one day in Yarmouth, Nova Scotia, in 1833. On that day, sailor and fisherman John Patch used a two-bladed, fan-shaped propeller to move a rowboat across the harbour. In so doing, the man and his creation effectively put an end to the age of sail. That Captain Patch's name is largely lost to history is a quirk of patent-office rules. At the time, Nova Scotia had no mechanism through which he could secure a patent for his propeller. He couldn't afford to travel to Great Britain to register his invention there, and U.S. patents were open to American citizens only. Yet as long as the winds shall blow, Canadians will remember that long before similar innovations in the United Kingdom and the United States, it was Patch's prop that ended an era.

ODOMETER
The details of distance.

Canadians, perhaps by necessity, have a fascination with distance, an understandable obsession in a land so vast with terrain so varied. No surprise, then, that a Canadian first developed the modern odometer. That Canadian is Samuel McKeen of Mabou, Nova Scotia. His device was made of a series of gear plates mounted on the frame of a horse-drawn carriage. The smallest plate in the series was engaged with a pinion on the hub of the carriage's wheel. As the wheel rotated, this plate became engaged and, as the carriage travelled, gradually engaged the increasingly larger gear plates in turn. Each plate rotated in sequence, ticking off a carefully calibrated amount of distance as it moved. An indicator hand encircled by a dial expressed the movements of the gear plates and therefore the distance travelled by the carriage. The innovation made it possible for the first time to assess how much farther a car could go with the fuel it had onboard. After that, there was no excuse for running out of gas. While today's odometers may sport glowing digital displays, they are not much different than Samuel's original built in Cape Breton in 1854. The carriages that transport us may have changed; how we gauge the details of distance has not.

THINK INNOVATION BEFORE INVENTION: HERE'S HOW.

☐ Think of innovation as an endless journey of creativity, not a destination.

☐ Decide why things are done or used *before* figuring out how to improve them.

☐ Define innovation as making something better to suit a particular purpose.

☐ Know that to have real impact, every innovation must be constantly improved.

☐ Be confident that people will welcome change when they understand how a change will make things better.

CANADIAN ABORIGINAL SYLLABICS
The signposts of meaning.

Borrow from the best. Advances often occur when people, without apology, take proven methods and apply them to new challenges. James Evans did just that when he created a new script for an ancient language. An Englishman, amateur linguist, and missionary living among the Cree in northern Manitoba in 1840, he adapted the Devangari script used in British India and British shorthand, which he had used as a merchant back home, to design a written language that reflected the oral traditions of the local Cree. Traditional Latin script isn't useful for most Aboriginal languages because they rely on long polysyllabic words. Evans's new script proved ideal. It was so easy to learn that literacy among the Cree was soon much greater than among French- and English-speaking Canadians. The new syllabic script was incorporated into Cree culture so quickly and spread so rapidly to other bands that subsequent newcomers to Canada assumed it had been in use for centuries. In the years since, other Aboriginal groups in Canada have taken advantage of the Evans script to create versions of their own. They too borrowed from the best.

OIL PIPELINE
The aqueduct of liquid energy.

In January 1862, Hugh Nixon Shaw drove a well near Petrolia, Ontario, looking for oil. Even he, no doubt, didn't expect what was to come. When the dig reached 160 feet, reports say, Mr. Shaw struck a gusher that pumped uncontrollably for a week, covering the nearby land in a layer a foot thick. Oil doesn't do anyone any good pooling up in fields, so the citizens of Canada's Victorian oil town got to work. Within a few years, they had built a pipeline to carry the oil twenty-five kilometres from Petrolia to Sarnia, where it could be refined. Not just any pipeline, mind you: the world's first.

STEAM BUGGY
The horseless carriage.

Decades before Henry Ford arrived on the scene, another Henry was hard at work. Henry Taylor of Stanstead, Quebec, built the first passenger automobile in North America — perhaps the world. He revealed his machine to the townsfolk in September 1868, when he drove it to the local fairgrounds for the annual farm festival. Henry's horseless carriage was just that — a two-passenger carriage — but instead of a horse up front, his conveyance was equipped with a coal-fired boiler on the back and a steam engine underneath. The Taylor steam buggy weighed 227 kilograms and could travel as fast as 24 kilometres per hour. Henry toured the countryside in his carriage for several years, showing it off at county fairs through the region. He even ventured across the border into Vermont. And then, it disappeared. Henry apparently grew tired of his steam buggy, dismantled it, and stored it in his attic — where it remained untouched and forgotten for the next hundred years.

TELEPHONE
The talking telegraph.

The telegraph seems like such an ancient device. Encoded messages made up of dots and dashes pulsing over a line. Yet in the 1870s, it was the world's most immediate communications tool, and innovators raced to uncover a way to squeeze more signals down a telegraph wire. Most approached the challenge by trying to get electricity to carry a range of sounds that imitated speech. One man saw the problem differently. Alexander Graham Bell set out to create an electronic appliance modelled on human physiology — the ear, to be exact. He wanted his creation to be an extension of the human being and not merely an enhancement of an imperfect device. This thinking stemmed from two facts: Bell was a speech pathologist and teacher of the deaf who had a keen under- standing of the human voice. He cultivated much of that knowledge while living in Brantford, Ontario, where he first began to study the human voice and experiment with sound in a workshop he called his "dreaming place." He also benefited from examining the complete structure of the human ear. (Don't worry, it came from a cadaver.) Only then did he appreci- ate how delicate its construction and sensitive its ability. After his 1874 innovation was unveiled, Bell himself summed up his advantage suc- cinctly: "Had I known more about electricity and less about sound I would never have invented the telephone." His talking telegraph remains meaningful, enabling today's world to reinvent continually how it communicates. Its voice reverberates across the years, as clear and brilliant as ever.

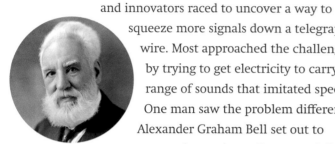

HANDSET
The practical telephone.

Alexander Graham Bell may have invented the telephone, but Cyrille Duquette made it a hands-on device. In 1878, the Quebec clockmaker, musician, and municipal councillor developed the first telephone handset. Still in use today around the world on landline telephones, the familiar handset joined the transmitter and receiver in a single handheld unit. Cyrille's name was mostly lost to history when he sold the rights to his creation for $2,100. He did so in despair after trying and failing to secure a loan for the development of his invention, which he hoped to forward by acquiring from Bell the rights to the telephone in Quebec City. Bankers determined that his handset had no future. Not quite. His handset was the very innovation that made the telephone practical. Sadly, lack of investment is still the bane of the innovator's existence, in Canada and around the world.

STANDARD TIME
The world's new clock.

Sandford Fleming was tired of showing up at the station for the nine o'clock only to find it had left at ten or twelve or fifteen minutes before the hour. Strange as it may seem today, there was a time when the actual time of day itself wasn't definite. Mr. Fleming's idea of nine o'clock wasn't necessarily the conductor's. One town's ten o'clock could be another's ten-fifteen or nine-thirty. Fleming brought order to the chaos. The polymath's standard time synchronized clocks within a region to a single time, using longitude to divide the world into twenty-four of these regions, or time zones. He proposed the idea in 1879, when a scholarly paper he had written about it was read to a gathering of the Royal Canadian Institute. His idea led to the International Prime Meridian Conference in Washington, D.C., in 1883, at which his system of standard time became the world's new clock.

TRANSCONTINENTAL RAILROAD
The steel link.

"All I can say is that the work has been done well in every way." William Cornelius Van Horne was being modest. Exceedingly so. When he uttered these words in 1885, the general manager of the Canadian Pacific Railway had just witnessed the driving of the last spike to complete Canada's first transcontinental railway. It was thought to be an impossibly ambitious undertaking. Foolhardy even. Just ten years earlier, then prime minister Alexander Mackenzie asserted that linking the port of Montreal to the Pacific Coast in the next decade could not be achieved "with all the power of men and all the money in the empire." Yet it was done—more than four thousand kilometres of steel laid across the land in five years, half the time allotted under the contract. Not just any land. The line blasted through the Canadian Shield and the world's oldest rock, spanned the sweeping plains, and went up and around and straight through the Rockies. Ingenuity was a daily requirement and

deliverable—to drain lakes and swamps, chop paths through bush and forests, fling bridges across rivers and gorges, drill tunnels into mountains. Courage, too, from each of the thirty thousand working men from across the country and around the world—many of whom gave their very lives to ensure the success of the undertaking. The combined effect of this ingenuity and courage was not only a railway but also a nation. The Canadian Pacific Railway fulfilled a condition of British Columbia's 1871 entry into Canada's Confederation. The railway also strengthened the country's claim to the remaining land of British North America not yet constituted as provinces and territories of Canada. In doing so, it served as an embankment of steel against any northward expression of American ambitions. Sir John A. Macdonald, prime minister at the time, put into words the enduring value of this supreme achievement of innovation and determination when he said, "We are made one people by that road…. That iron link has bound us together in such a way that we stand superior to most of the shafts of ill fortune."

RADIO VOICE TRANSMISSION
The everywhere sound.

It was an encounter between a man recognized as the world's greatest inventor and a boy who would become his equal. The year was 1876 and Reginald Fessenden's physics-teacher uncle took the young lad from East Bolton, Quebec, to meet Alexander Graham Bell soon after Mr. Bell had developed his signature instrument. Fessenden was impressed by the device but wondered why it needed wires to connect two of the sets. Couldn't – shouldn't – voices be able to travel through the air without wires? Twenty-four years later, he had the answer. Fessenden's first speech broadcast took place on December 23, 1900, when he transmitted his own voice over the first wireless telephone from a site on Cobb Island in the middle of the Potomac River near Washington, D.C. This was radio, a continuous wave of sound sent out from a transmitting station. And it happened a full year before Guglielmo Marconi's fragmented transatlantic transmission of dots and dashes. The unassuming Canadian pressed on. Six years later, he broadcast music and speech to ships plying the Atlantic and achieved two-way voice communication between Massachusetts and Scotland. And what did this genius get for his miraculous achievements? Not fame – Marconi was considered the darling of radio even though his accomplishments paled in comparison to Fessenden's. Not riches – his backers seized his patents and pushed him aside, forcing him to spend years and thousands in court. Not respect – not only did his home country of Canada not supply him with financial support to carry out further research but it also refused his request to create a radio network, a privilege granted to Marconi instead. Despite the disappointments and setbacks, Fessenden carried on. Did he ever. Over the course of his career, he developed more than a hundred patentable inventions, including something he created in 1929 – television. Ever heard of it? Fessenden: his name may not be well known in our day but his ideas are all around us.

NATIONAL ATLAS
The first book of country maps.

Canada is one of the most diverse countries in the world. It deserves a book that does justice to that diversity. First published in 1905 by the country's department of the interior, the *National Atlas of Canada* is that book. The world's first

national atlas, the book was made up of a series of thematic maps that describe the country's geology, population, natural resources, communications, and economic activities. The sixth edition of the atlas was launched on the Internet, becoming the world's first electronic online national atlas. The most recent digital version takes full advantage of the medium, relying on multimedia resources and interactive graphics to tell the evolving story of an increasingly diverse nation. Always good to know where you come from.

SILVER DART
The first team-built aircraft.

Aerial Experimental Association was an apt moniker. The association was just that: a group of men, including famed father of the telephone Alexander Graham Bell. They were engaged in an experiment. Not just any experiment. In 1909, they built the first powered aircraft to fly in Canada. In Baddeck, Nova Scotia, to be exact. Its name was the Silver Dart. The aircraft, named for the silver-coloured rubberized fabric used to cover it, flew on its second attempt. Taking off from the ice of Baddeck Bay, it travelled 800 metres at an elevation of up to 9 metres at roughly 65 kilometres per hour. While not the first powered aircraft to fly, it was the first built by a team. Its leader was the aforementioned Mr. Bell. Glenn Curtis supplied its engine, Thomas Baldwin its silver fabric, and Frederick Baldwin and John McCurdy its design. Young McCurdy, twenty-three years old at the time, also doubled as pilot. Even teamwork has its limits.

CURTISS CANUCK
The barnstormer's choice.

In 1917, when Canada needed a plane to train pilots for the fledgling Royal Flying Corps, they turned to the American Curtiss JN-4. That aircraft was a biplane with an upper-wing aileron—a hinged surface at the end of a wing that can be raised or lowered to make an aircraft bank and roll. To make sure their aircraft met the exacting specifications of flight training, Canadian officials modified the biplane so it was equipped with ailerons on the bottom wing as well as the top. This move gave the Curtiss Canuck better lateral control than its U.S. cousin and spurred mass production of the aircraft with this feature to meet demand not only in Canada, but also for pilot training in the United States. The modification also turned the aircraft into a post-war favourite among flyers of all kinds—aerial surveyors, airmail pilots, and ski flyers. Even barnstorming stuntmen, many of them starting out as World War One pilots and returning, appropriately enough, full circle.

MOUNT LOGAN BAROMETER
The height of practicality.

New tasks often require new tools. In 1925, the Alpine Club of Canada organized a three-country expedition to summit Mount Logan in the Yukon for the first time. Known as North America's Everest, Mount Logan made for a formidable ascent. It's the highest mountain in Canada and is believed to have the largest circumference of any mountain on earth. While the men of the expedition were up to the harsh conditions of the challenge, no existing barometer was. So the Dominion Land Survey created a new one. The organization's device—the Mount Logan barometer—is an aneroid barometer. The instrument measures atmospheric pressure at a given altitude through the pressure's effect on an aneroid, a metal chamber partially evacuated of air. H.F. Lambert of Canada's Geodetic Survey used the Mount Logan barometer all the way to the top. A new device to reach a new summit.

DOCUMENTARY FILM
The storyteller's new canvas.

Nearly a century has passed since it was made and it remains one of the most vital films ever produced – vital in its absolute necessity in explaining the history of filmmaking, and vital in its depiction of one of the most unforgettable figures ever to appear on screen. The film is *Nanook of the North.* Produced by Robert Flaherty – an explorer, prospector, and fledgling filmmaker (like all filmmakers at the time) – it is widely recognized as the first feature-length documentary film. Released in 1922, the film follows the lives of an Inuk man named Nanook and his family, pictured here in a still from the documentary, as they travel, trade, and hunt in the Canadian Arctic.

Some criticized the Quebec filmmaker for staging several of the scenes. Others came to his defence, arguing that the technologies of the time made some staging inescapable. Flaherty himself argued that documentary filmmakers – a new term for a new art – must sometimes distort an authentic thing or event to capture its true spirit. Regardless of the controversy, audiences fell in love with the story and the medium. Part of the film's appeal was its depiction of an unknown world whose spirit of innovation, community, self-preservation, peacefulness, and resourcefulness was of huge interest to Western society – a world that had just come through the most brutal war in history. The story of Nanook and the Inuit gave hope that all humans could be better people. One hundred years later, we still turn to stories on film – both real and fictional – to gain a greater understanding of the best in the human spirit.

IMPROVE A PROCESS: HERE'S HOW.

☐ Identify any series of steps by which something is done repetitively.

☐ Clearly define the *ultimate goal* of that process.

☐ Break the process down into its single steps, and identify who is responsible for each.

☐ Remove any step that does not contribute to the ultimate goal of the process.

☐ Add any other step that helps reach the goal faster or better.

☐ Reassign any step that could be better done by someone else.

☐ Automate any process only when its outcome is perfectly consistent and desirable.

STREAMLINED LOCOMOTIVE
The windjammer on rails.

Wind tunnels aren't just for cars and planes. Decades ago, Canadian researchers were using them to design locomotives. In 1930, engineers — the mechanical kind, not the train-driving kind — at Canada's National Research Council used the organization's new wind tunnel to test current models of locomotives and experiment with new designs. Their studies led to the launch of streamlined locomotives. These sleek machines not only cut through the wind, but also didn't foul the vision of engineers — the train-driving kind this time — with coal and diesel smoke. Soon railways were using the findings to produce ever more trim and elegant locomotives. A new era in transportation design was on the rails.

STARS IN GLOBULAR CLUSTERS
The new way to see the universe.

Helen Battles Sawyer Hogg is the greatest involuntary volunteer in the history of astronomy. Dr. Hogg carried out pioneering research into globular clusters and variable stars. (Globular clusters are spherical collections of stars that orbit galactic cores as satellites. Variable stars are stars whose brightness, as seen from earth, fluctuates.) She did her groundbreaking work while serving as a volunteer assistant to her husband while he worked first at the Dominion Astrophysical Observatory in Victoria, British Columbia, and then at the David Dunlap Observatory at the University of Toronto. Neither place would hire a woman — even one who had earned a doctorate — let alone a married one. While in Victoria, Dr. Hogg began taking photos of variable stars, cataloguing the cyclical changes in their brightness. In the process of doing so, she uncovered 132 new variable stars in the global cluster Messier. She continued her work in Toronto, amassing thousands of photographs of globular clusters that she used to identify many thousands of variable stars. In 1939, she published the first of three massive catalogues of variable stars in globular clusters — publications that are still used widely today. She also used her deep understanding of variable stars to enhance our knowledge of the age, size, and structure of the Milky Way. And for some thirty years, Dr. Hogg wrote a weekly astronomy column for the *Toronto Star*, sharing her wisdom, passion, and curiosity with more than a generation of Canadians. All in all, pretty good work for a volunteer.

SNOWPLANE
The magic carpet of the Prairies.

Living in such a vast and sparsely populated land, Canadians have always looked for ways to bridge the great physical distances that separate them. Nowhere are these distances more pronounced than in remote towns on the Prairies during the long Canadian winter. In 1935, Karl Lorch devised a novel way to conquer isolation. The Spy Hill, Saskatchewan, resident created the world's first snowplane. He built the cabin structure of his lightweight machine using old aircraft tubes covered with treated linen. He powered his machine with an old six-cylinder aircraft engine and balanced the whole thing on three skis. Karl's snowplane went from oddity to celebrity; orders for the machine came quickly from doctors, taxi drivers, government inspectors, power-line repairers, and telephone installers — anyone who needed to get around in the snow. Even the Canadian army was a buyer. They were all drawn to the Lorch snowplane's speed and superiority over track snowmobiles in navigating deep snow. The snowplane proved a reliable way to overcome the isolation that was so much a part of Prairie life.

SNOWMOBILE
The winter bus.

Armand Bombardier was just a regular repair-man. Living in the small Quebec town of Valcourt, just east of Montreal, he fixed cars, tinkered with machines, and sold gasoline. His world changed forever on a stormy winter night in 1934. Most streets and roads in the town weren't ploughed back then. When a blizzard roared through Valcourt, Armand couldn't transport his son, stricken with a sudden attack of appendicitis, to the local hospital. The young boy died. The family tragedy stirred Armand to build a vehicle that could travel quickly and reliably in even the snowiest, iciest conditions. He introduced his first snowmobile in 1937. Made to carry seven people, its main feature was a sprocket wheel and track-drive system that gave the vehicle grip and power to race across snowy streets and fields. The machine became popular immediately with professionals – from priests to postmen to – ironically enough – ambulance drivers. Out of one family's tragedy, a Canadian transportation legend was born.

WALKIE-TALKIE
The point-to-point communicator.

Donald Hings probably had little idea just how useful his creation was about to become. In 1937, the Canadian inventor created the first handheld portable two-way radio transceiver. Donald called it the packset. It soon gained widespread use and fame under a more descriptive name – the walkie-talkie. Within two years, war erupted in Europe and Donald was summoned to Ottawa to adapt his packset for military use. Thousands of the sturdy and reliable devices were soon in the hands of Allied infantrymen around the world. They became even more popular after the war. First responders relied on them as essential equipment. Truck drivers used them to report emergencies and stay in touch while on the road. Even kids were equipped with them, roaming around their neighbourhoods during the day and chatting under the covers at night when they should have been sleeping. 10-4. Roger out.

AIR AMBULANCE
The wings of mercy.

What is it about Saskatchewan? The province seems to be the birthplace of a disproportionately large number of advances in the delivery of health care. Add the Saskatchewan Air Ambulance Service—the first non-military, government-operated air ambulance service in the world—to the list. Taking off in 1946, the service's one plane, one pilot, one nurse, and one engineer made sure people in remote areas of the province who were victims of accidents or suffered from acute illnesses received quality medical attention. Within a handful of years, the province's fleet of air ambulances were ferrying more than a thousand patients each year—many of them suffering from polio—to clinics and hospitals. The air ambulance is so commonplace around the world today that the idea seems self-evident, wings of mercy carrying those in distress to the care they need.

DE HAVILLAND BEAVER
The workhorse of the bush league.

Much of Canada is accessible only by plane. And not just any plane. It must be able to carry heavy loads and take off and touch down on short strips of land or water. The Beaver is made for these rugged, often inhospitable conditions. Designed in 1949 by R.D. Hiscocks of De Havilland Aircraft in Toronto, the aircraft's all-metal construction and high-lift wing and flap configuration make it a sturdy, reliable workhorse of the northland. Not that it was confined to Canada. The appeal of the Beaver quickly became apparent to others. Buyers of all types emerged in some sixty countries around the world. The United States Air Force relied on them during the Korean War to ferry men and materiel, while Sir Edmund Hillary and his team used the Beaver in their mission to the South Pole. From northland to South Pole, this enterprising beaver has seen and done it all.

ORENDA
The most powerful jet engine.

As World War Two ground on, countries on both sides raced to develop technologies that would tip the balance in their favour. One of those technologies was the jet engine. In 1942, Canadian researchers set up a facility to test early versions and determine how they performed in cold weather, a quintessential Canadian concern. This work enabled them to design and develop many new jet-engine technologies. By 1949, three researchers — K.F. Tupper, Paul Dilworth, and Winnett Boyd — had set up their own factory in Malton, just north of Toronto. The factory was the hub of a new company — Avro Canada — and the birthplace of what was then the world's most powerful and reliable jet engine. Its name was Orenda. The company manufactured some four thousand Orenda engines, supplying them to air forces throughout the world, including that of West Germany. The country that had spurred the Canadian effort to build the best jet engine ended up becoming the beneficiary of that effort.

ALOUETTE SATELLITE
The atmospheric wonder.

Alouette 1 was built for a mission that was to last one year. Ten years after being launched in 1962, Canada's first satellite was still hard at work taking and transmitting what would turn out to be more than one million images of the earth's ionosphere. Extending from about eighty to a thousand kilometres above the earth's surface, the ionosphere is a layer of the planet's atmosphere with a high concentration of ions and free electrons and is able to reflect radio waves. Engineers working in Canada's Defence and Research Telecommunications Establishment just outside Ottawa built Alouette 1. It was an orbital workhorse — made up of an ionospheric sounder, VLF (very low frequency) receiver, and energetic particle detector — that gave the world its first top-down understanding of the ionosphere. In the process, the one-year wonder that lasted a decade brought us fresh insights into the makeup of the planet's atmosphere and led to many improvements in the way we communicate back here at home.

STEM ANTENNA
The big bounce.

Successfully exploring new frontiers is the truest test of ingenuity. George Klein showed he was up to the challenge of the final frontier. In 1952, the mechanical engineer at Canada's National Research Council was asked to create an antenna for Canada's first satellites. Alouette I and II were being readied to study the ionosphere by bouncing radio signals off that charged layer of the earth's atmosphere from above. The satellite would need not only several antennas to carry out the big bounce but also ones solid and compact enough to withstand the punishing effects of space travel. So George, with a broad team of experts working hand in hand, produced the STEM antenna. Its name an acronym for storable tubular extendible member, the antenna looks like a roll of coiled steel or a large measuring tape. A small motor extends the antenna until it takes the form of a rigid tube. Early versions were six metres long and in time increased to become as long as an amazing forty-five metres. The device could also be improvised to suit a range of applications – from serving as legs on survey equipment to moving equipment inside nuclear facilities. Whatever the size, whatever the task, the STEM antenna got the job done and became a welcome fixture on the famed Mercury, Gemini, and Apollo missions.

SKI-DOO
The typo that became a sport.

The end of World War Two dealt Armand Bombardier's snowmobile company a double whammy. The first was obvious and expected: contracts to supply Allied governments with specialized military vehicles halted abruptly. The second wasn't anticipated: governments in Quebec – both provincial and municipal – began clearing snow from roads in wintertime. Ploughed streets and highways throughout the province meant professionals who had once relied on snowmobiles to go from place to place could now use their automobiles year-round. With two dependable markets gone in virtually a flash, Armand set about to create a new one. Starting in the 1950s, his company built smaller snowmobiles meant for one or two passengers and for one purpose – fun. He wanted to call his new recreational vehicles the Ski-Dog. A typo caused the name to appear as Ski-Doo. Armand preferred the mistake. So did fun-loving people throughout the province and across the country. By the mid-1960s, some 8,500 Ski-Doos were being sold each year. Build it and they will ride.

EURO
The regional currency.

Sometimes the fastest path to innovation comes in casting aside played-out assumptions. Robert Mundell of Kingston, Ontario, was one for discarding outdated thinking to get at something new: if states within countries could share the same currency, he thought, what stopped countries within regions from doing the same? Some speculate that his being from Canada — a country not only made up of many disparate parts but also whose economic welfare had been tied to the fortunes of more powerful nations for generations — inspired his proposition. The economist's freeing premise led him to develop the concept of a regional currency in 1961. His idea took on even greater, ahem, currency as the post-war world experienced rapidly increasing levels of international trade and capital movement. Dr. Mundell believed a shared currency, in this new kind of world especially, could enable countries within regions to maximize their economic efficiency. His carefully articulated notion paved the way for the fourteen founding jurisdictions of the Eurozone to adopt the Euro on January 1, 1999, the same year Mundell accepted the Nobel Prize in economics. Some forty years after Mundell first articulated the idea, the regional currency became real.

MEDIA STUDIES
The mind of Marshall McLuhan.

There was a time when little if any thought was given to the impact of communications technologies beyond where and how much each was used. Marshall McLuhan changed all that. In the 1960s, the University of Toronto English professor showed the world how throughout human history individual technologies have shaped societies, cultures, and people themselves. Even those who have never heard the name McLuhan probably quote him. His phrases are fixtures in the popular vernacular of the Western world. He spoke of our world as a "global village" made ever smaller by the rapid movement of information. He claimed "the medium is the message," because the medium through which a message travels determines the scale and form of human association and action. And he said, "We shape our tools and then our tools shape us," noting that communications technologies are extensions of ourselves and not things apart. The insights behind these phrases ushered in the international discipline of media studies, a field of immediate value as we come to grips with the unique characteristics of our own digital age.

AVRO ARROW

The child of the wind tunnel.

Canada's response to the threat of Soviet nuclear bombers was an arrow. Not just any arrow. The Avro Arrow—designed and developed by A.V. Roe Canada—was ahead of its time technologically. A marvel of engineering, it was a twin-engine, all-weather supersonic interceptor with a computerized control system that enabled pilots to control the craft electronically rather than manually. Although the Arrow was ahead of its time conceptually, it quickly fell behind the time strategically. In 1958, the year the Arrow first left the quiver and took flight, Soviet missiles had replaced Soviet bombers, making jet interceptors obsolete almost overnight. Within a year, the Arrow was grounded forever.

DIGITAL TELEPHONE SWITCH
The new central office.

Digital World. It's a fitting title for the world's first line of fully digital telecommunications products. Developed by Nortel Networks engineers in Mississauga, Ontario, in 1975, Digital World featured DMS-100, a central office digital telephone switch that could serve up to 100,000 lines. Unlike all previous switches, the DMS-100 was fully digital and therefore fast enough to handle (and perhaps enable) an exponential growth in telephone traffic. The advance was urgently needed and therefore perfectly timed. It soon became the backbone of telephone systems around the world and set the stage for the world's first standard-based commercial application of packet switching, the foundation of the Internet. Thanks to a Canadian team of electrical engineers, the telecommunications central office was now fully digital.

CANADARM
The Canadian handshake.

What's better than one Canadarm? The answer could only be two. Developed for NASA by a team of engineers from Canada's SPAR Aerospace, the first Canadarm was a marvel of engineering that meets exacting criteria for weight, versatility, safety, reliability, and precision of movement. Deployed in 1981, it proved an essential component of the U.S. shuttle program for thirty years. Astronauts used it to maintain equipment, deploy and capture satellites, and move around all manner of cargo and payloads — even their fellow astronauts. The program eventually relied on five versions of the original Canadarm. A second Canadarm is now hard at work on the International Space Station, doing much the same as the original performed for the shuttle. In tandem, the two arms partnered in constructing the station, the shuttle-borne Canadarm handing over segments of the station for assembly by the station's arm in a move quickly dubbed *the Canadian handshake*. The Canadarm — and engineering genius behind it — is also hard at work at home. Canadian space technology is being relied on to develop a series of tools to make surgeries safer, more accurate, and less invasive. The Canadian handshake lives on.

LANGUAGE THEORY
The grammar instinct.

Human beings have an instinct to speak. According to Steven Pinker, language is a behaviour unique to human beings shaped by natural selection to solve the specific problem of communication inherent to our hunter-gatherer species. It is as instinctual to us as is dam building to beavers and web weaving to spiders. This is the groundbreaking thesis found in the Montreal-born experimental psychologist's 1989 book, *The Language Instinct*. To arrive at this science of language, Dr. Pinker combined for the first time cognitive science, behavioural genetics, and evolutionary psychology. Along with constructing this new understanding, Dr. Pinker's insights overturn many widely held beliefs about language — that it must be taught, that people's grammar is growing worse as new ways of speaking emerge, that language limits the kinds of thoughts a person can have, and that other great apes can learn languages. Human beings have an instinct to speak and, for better or worse, it is ours alone.

SPACE VISION
The secret of the black dots.

Inspiration can emerge from the unlikeliest places. Dr. Lloyd Pinkney's idea for space vision came while watching a movie. In 1992, the engineer at Canada's National Research Council saw a character run past a wall and it hit him: he would be able to measure the character's movement precisely if he added fixed points to him. Dr. Pinkney's revelation is based on the principle that our brains judge the location, orientation, and speed of objects best when their position is made relative to something else. Yet it's tough to do that in the nearly black void of space, where reference points are few. Back at the lab, Dr. Pinkney worked with a team of colleagues on a system of space vision in which black dots would be affixed to the items in the Space Shuttle's payload. Cameras attached to the Canadarm would capture sequential images of each item every thirty-three milliseconds. Using the dots as reference points, the Space Vision System would calculate the location, orientation, and speed at which a piece of payload was moving relative to the giant arm. Shuttle astronauts who controlled the Canadarm would then use that data to approach, seize, and assemble payload items safely and reliably. The Space Vision System soon became a fixture on shuttle flights. It was instrumental in enabling astronauts to connect the first two elements of the International Space Station in 1998. In 2003, the system was updated to carry out in-orbit examinations of the shuttle itself. It's a secret that keeps being revealed.

56K MODEM
The direct-dial appliance.

Brent Townshend was just trying to find a better way to download music. In 1993, while working on a system he was building called Music Fax, the Toronto audiophile found that download speeds from servers connected to the phone network through digital links could reach 56 kilobits per second because the downloaded files didn't have to be converted from analog to digital. He soon patented this technology, establishing the 56K modem as the industry standard for accessing the Internet through analog phone lines. So long, Music Fax. Hello, world.

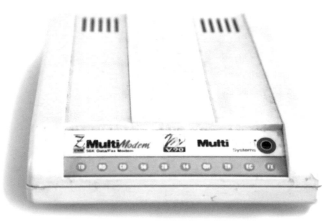

JAVA
The programming language.

"Write an application that will run everywhere." That mission doesn't sound impossible today, but twenty-five years ago it was an incitement to revolution, and the very challenge Calgary's James Gosling and his team of Sun Microsystems engineers set out to conquer. In 1995, Gosling presented their answer: Java. Before Java, programming languages were complex, tedious, and limited largely to desktops and mainframe computers; handheld and imbedded devices were still impractical dreams. Java was designed so its compiled code could run on all platforms without having to be recompiled. Equipped with Java, programmers were at last free to imagine all manner of new devices – especially smaller and specialized ones. Notably, Java made it possible for computer programmers to take advantage of the emerging web era, making the user experience platform-independent while speeding its development. Today, Java remains the primary language in which to develop and deliver content to the web. It is the nimble intelligence that brings ten billion digital devices to life. A quarter century after its innovation, Gosling's Java is *the* programming language.

TWO-WAY MESSAGING
The packet conversation.

Think outside the box. Here's an example of that classic innovator's rule in action. In the 1990s, several high-tech companies had developed prototype two-way mobile messaging devices, yet none of them could crack the one big problem: how to send data efficiently, affordably, and securely over a wireless network. Mike Lazaridis found the answer by looking in a whole different direction. First, the Waterloo engineer and his colleagues at Research In Motion compressed and separated their data into packets. Then they created software programs so this information could be received and understood by computers. This packet approach to digital communication made it possible by 1996 for people to use their mobile devices to retrieve emails from their computer desktops. Second, even though they had figured out how to send data efficiently, the cost of using the cell phone network for these transmissions was prohibitive. Lazaridis responded by sneaking his new tech onto the sleepy pager network — a lesser-used and much cheaper part of the radiowave spectrum auctioned to carriers by national governments. It worked. The network was affordable and secure. Research In Motion eventually used this same medium for its BlackBerry Messenger service. Inside the old network was outside the box.

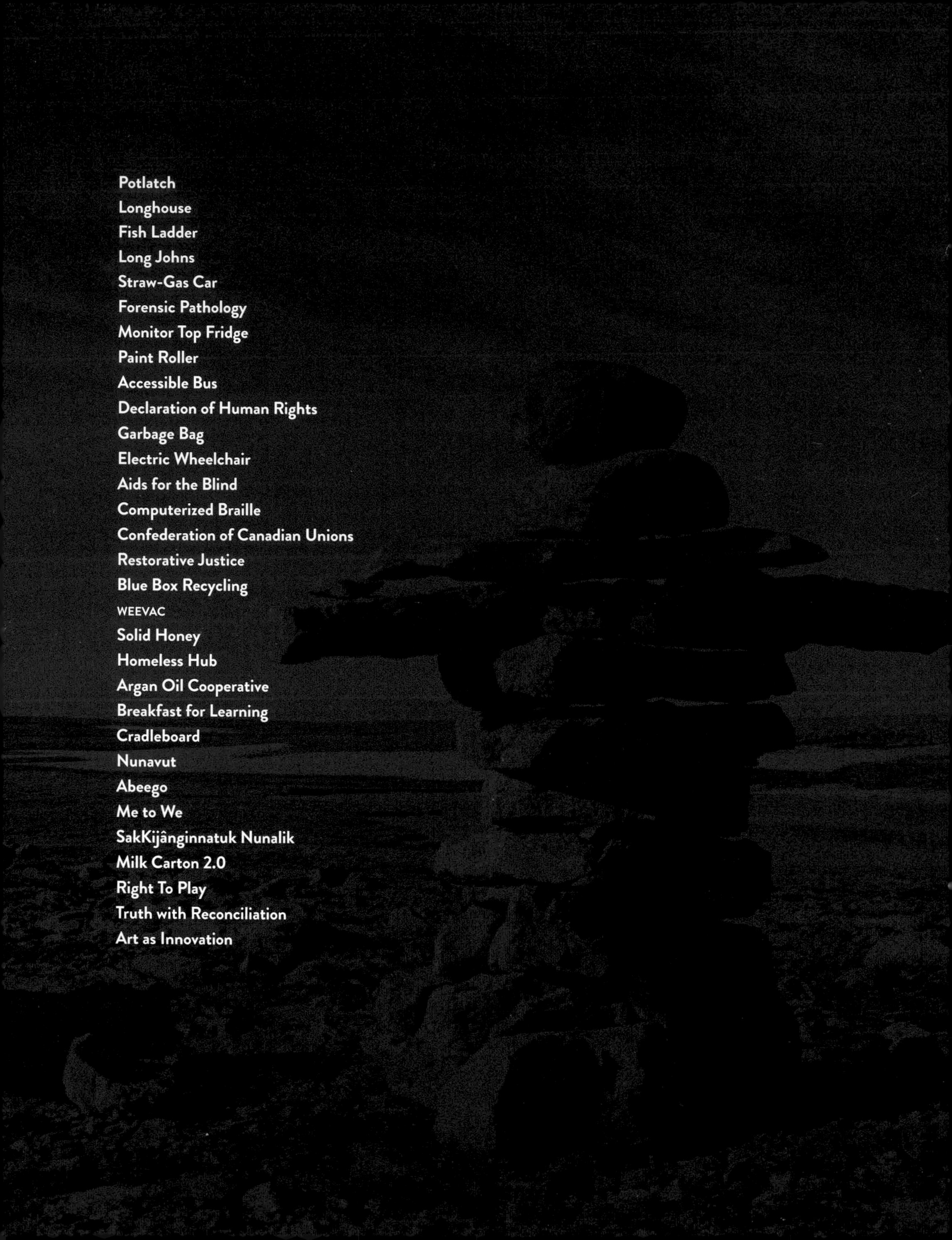

Potlatch
Longhouse
Fish Ladder
Long Johns
Straw-Gas Car
Forensic Pathology
Monitor Top Fridge
Paint Roller
Accessible Bus
Declaration of Human Rights
Garbage Bag
Electric Wheelchair
Aids for the Blind
Computerized Braille
Confederation of Canadian Unions
Restorative Justice
Blue Box Recycling
WEEVAC
Solid Honey
Homeless Hub
Argan Oil Cooperative
Breakfast for Learning
Cradleboard
Nunavut
Abeego
Me to We
SakKijânginnatuk Nunalik
Milk Carton 2.0
Right To Play
Truth with Reconciliation
Art as Innovation

THREE

Smarter

Smaller

Kinder

Safer

Being fair matters to Canadians. They speak openly about their desire for a country that is both smart and caring. In the desire to find better ways of treating each other, Canadians have devised innovations that have quietly made the world a kinder place. They have given the world novel ways of distributing wealth, keeping the peace, righting wrongs, accommodating the disabled, fostering pride, moving victims to safety, and protecting human rights. It's good to be kind.

Healthier

Wealthier

Happier

POTLATCH
The system of sharing.

For the Indigenous peoples of Canada's Pacific coast, status has never been displayed by the accumulation of wealth. On the contrary, for thousands of years they earned community respect by giving away goods, holding elaborate ceremonies of feasting and dancing at which the giving took place. Known as potlatches, these gatherings were held most often in winter months and usually coincided with births, deaths, marriages, and other important occasions. Held among members of a community or between members of two, potlatches served several valuable purposes. They were primarily economic instruments that enabled families in need to procure wealth such as food and animal skins, which they then brought home to share with friends and neighbours. These were also rituals at which political, religious, and kinship titles and roles were transferred between people, bestowing increased status on the recipients. And they were opportunities for community leaders to showcase visibly and plainly their wealth and prominence through giving. In 1884, the Canadian government – to its everlasting shame – outlawed potlatches in an effort to speed the assimilation of the country's Aboriginal peoples. Government officials saw these ceremonies as being "contrary to civilized values of accumulation." On the "contrary," all cultures then and now have much to learn from this time-honoured system of sharing.

LONGHOUSE
The cosy community.

Communal living is experiencing a renaissance in many parts of the developed world. Not apartment buildings or condo towers but shared living spaces that accommodate extended families and take maximum advantage of available room. The Iroquois and Huron people of Canada and the northern United States had this kind of living figured out thousands of years ago. Longhouses were central living spaces in Aboriginal communities, especially among the Iroquois peoples of what is today Quebec. Iroquois longhouses were singular struc-

tures that ranged in length from 30 to several hundred feet. Typical ones were some 200 feet long, 20 feet wide, and 20 feet high. Each longhouse was divided into compartments, with an aisle running down the centre of the entire building. Two families lived in each compartment, and everyone slept in bunk beds, another Iroquois innovation now used worldwide. Roofs and ends were usually rounded — no sharp corners so little wasted space. And those ends could be extended to make room for expanding families. Longhouses did more than put a roof over people's heads. They were the structural expression of the deep community spirit of the Iroquois, Huron, and others. The buildings in which they lived were a reflection of how they must live in all aspects of their existence if they were to survive. The cosy community of the longhouse was not just comfortable; it was a means of survival and a way of life.

FISH LADDER
The salmon helper.

Human civilization, with only the briefest of pauses to reflect on the influence to the natural world of its methods and actions, has changed the habitat of countless species. There is at least one notable example of our effort to mend the fabric of the environment. The year was 1837. It began when Bathurst, New Brunswick, mill owner Richard McFarlan built a dam to harness the strength of the river's flow to power his lumber mill. To make it possible for salmon to return to their spawning grounds, Richard put together a fish ladder. Made up of a series of step-like ponds with connecting underwater passages, the structure enabled the salmon to ascend gradually around the dam. As people built bigger dams and other obstructions, fish ladders became larger and more sophisticated as well. Yet their basic idea remained true to McFarlan's original salmon helper.

LONG JOHNS
The warm, comfortable pair.

Thermal underwear used to come only in one piece that covered you from neck to ankles. Yet there were two problems with the one-piece. It was a lot of fabric to wash when only small parts of it needed cleaning. And not all bodies are created equal; even two people the same height might vary in the lengths of their legs and trunks. So for some, the one-piece was a bad fit. Enter Frank Stanfield. One piece of the famed Stanfield brothers underwear empire, Frank came up with the adjustable underwear combination in 1915. It was a simple innovation. Inserting one button from a vertical row on the shirt into one of many buttonholes on the pants, a wearer could build a pair to fit him or her perfectly. More popularly known as long johns, Frank's new undergarment kept all sizes both comfortable and warm; and comfort and warmth is a pair that fits all.

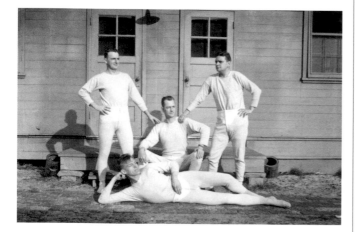

STRAW-GAS CAR
The cure for wartime fuel shortage.

Scarcity often fuels innovation. It certainly fuelled the straw-gas car. Not literally; straw really fuelled this vehicle. In 1917, University of Saskatchewan chemistry professor R.D. MacLaurin and his engineering colleague A.R. Greig were looking for a way to deal with the gasoline shortage caused by World War One. There was no shortage of straw, so they explored how to use the gaseous vapour produced by heating straw as fuel for vehicle engines. They started by heating baled straw in a retort, a large glass container with a long neck. They captured the gas, mostly methane, and used it to run a motor. To take the creation from the lab to the road, the two inventors rigged a large balloon filled with the straw gas to the frame of a car with a pipe leading from the balloon to the car's carburetor. Using a simple valve, the driver could switch from gasoline to straw gas and back as desired. War's end also brought an end to this curious-looking car. Yet what's old is new again. Intrepid innovators around the world are exploring how to use a range of alternative fuels to power vehicles. One hundred years later, who knows? Maybe straw will once again take us down a new road.

FORENSIC PATHOLOGY
The Sherlock of Saskatchewan.

Modern crime scene investigation methods did not appear first in Miami, New York City, Las Vegas, or even Scotland Yard. The originator of forensic pathology is a remarkable woman who worked on the Canadian Prairie. Dr. Frances McGill was the first person in the world to make this science a regular part of police investigations. Appointed Saskatchewan's chief pathologist in 1920, Dr. McGill travelled by any means necessary including dogsled and floatplane throughout the vast province to investigate suspicious deaths. Known as the Sherlock Holmes of Saskatchewan, she applied her training as a medical doctor to study crime scenes and protect and preserve evidence in ways that had never before been done. She also became renowned for her court-room appearances, where her riveting and rigorous testimony – anchored in medical science – would exonerate the innocent and convict the guilty. And she taught what she knew and had learned – how to tell human blood from animal blood, for instance – to students at the Regina Police Academy. Today, those insights and methods either are still in use or have inspired others and are relied on by police departments across Canada and around the world. Even Miami.

MONITOR TOP FRIDGE
The all-in-one icebox.

The iceman cometh no more. The first consumer refrigerator to integrate a refrigeration unit and cabinet was developed by General Electric in Toronto in 1927. The cooling unit was a cold-air compressor mounted conspicuously on the top of the cabinet. It was dubbed the *monitor* because the compressor looked like the turret of the USS *Monitor*, a celebrated ironclad warship launched during the United States Civil War. The monitor top fridge was a landmark advance in consumer appliances, giving families a reliable alternative to their legacy sawdust-filled iceboxes. Even so, the Canadian innovation didn't become popular until after World War Two, when electrification was commonplace and when higher wages and lower prices made them affordable for most families.

PAINT ROLLER
The do-it-yourselfer's blessing.

Do-it-yourselfers around the world, take a
moment of silence to honour Norman Breakey.
In 1939, he created the first paint roller. Great
idea. The roller applied paint evenly and made
painting itself faster than ever. Yet the visionary
Torontonian didn't patent the innovation – not
the fabric-covered cylinder, nor the long pole
shaped liked the number seven, nor the ridged
pan made from tin. Big mistake. Knock-offs
emerged quickly and others secured lucrative
patents by making minor modifications to
Norman's original idea. And what of Norman
himself? Not much if anything is known of his
fate. So before you paint that wall or ceiling, take
a moment to remember him. No one else will.

ACCESSIBLE BUS
The great enabler.

It's fine to say that all people should have equal
opportunities, but what are we doing to make
sure they actually do? Walter Callow did something. The Nova Scotian set up a wheelchair
coach service for disabled veterans in 1947. He
started the service by having two custom-made
buses built in nearby Pubnico, then convinced
the big boys at Ford and General Motors to
manufacture the wheelchair coaches on their
assembly lines. Today, municipal transportation
systems throughout the world include wheelchair-
accessible vehicles in their fleets. Ironically
enough, the only ride the blind, quadriplegic
Walter took in one of his coaches was when his
own body was transported to his funeral – a
service conducted with full military honours for
a man whose accessible bus was the great
enabler for thousands of veterans.

DECLARATION OF HUMAN RIGHTS
The precursor to peace on earth.

McGill University professor of law John Humphrey believed "there will be peace on earth when the rights of all are respected." He backed up this profound and noble sentiment with the best attempt made so far to identify the inalienable rights of all. Professor Humphrey's achievement is the *Universal Declaration of Human Rights* — a preamble and thirty brief clauses that encapsulate the rights to which each and every human being is entitled. He wrote the

declaration in his capacity as first director of the United Nations division of human rights. The then-fledgling international organization adopted the Canadian's moral and intellectual handiwork in 1948. Two-thirds of a century has passed and Humphrey's formal statement on behalf of the world remains Canada's greatest contribution to international law and the precursor to peace on earth.

GARBAGE BAG
The better bin.

It was the best of products, it was the worst of products; it was the age of the garbage bag. Sorry, Mr. Dickens, but that corrupted version of your famous phrase sums up the life of the plastic garbage bag. It was the brainchild of Harry Wasylyk, who made the first version in 1950 in his Winnipeg kitchen. Handy, clean, and surprisingly strong, the garbage bag made for a marked improvement over the clunky, dirty, smelly metal pail. Harry focused his sales on large commercial operations, and his first buyer was the Winnipeg General Hospital. But sales really took off when Union Carbide bought the invention and sold Harry's creation under the name Glad Garbage Bags. Variations of the garbage bag found their way into homes around the world. Now comes the worst part: by the 1970s, millions – maybe even billions – of virtually indestructible garbage bags were clogging landfills the world over. Enter James Guillet. This Canadian scientist came up with a degradable plastic that decomposes in the sun. Consider it a better plastic for the better bin.

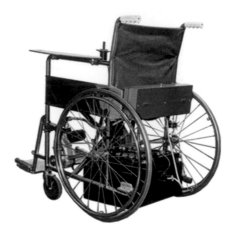

ELECTRIC WHEELCHAIR
The veteran's new legs.

War is often an exercise in unintended consequences. The wonder-drug penicillin, for instance, enabled thousands of gravely injured World War Two servicemen to survive their wounds, yet many of these otherwise doomed veterans returned to their homes and families as paraplegics and quadriplegics. Conventional wheelchairs were of little use to these men, whose manual strength and dexterity had been impaired or eliminated. George Klein embraced this new challenge – an unforeseen consequence of survival by Canadian vets. In 1953, the engineer at Canada's National Research Council developed the world's first electric wheelchair. Three features set it apart from anything to date: the voltage of the chair's motor enabled it to climb all but the steepest inclines; its power supply was strong enough for the chair to be used both indoors and outdoors over long stretches of time; and independent drives for each of its wheels gave the chair the ability to pivot sharply and thereby operate in close quarters. The control switch on this nimble wheelchair resembled the joystick found on today's game consoles. George worked closely with patients to adjust the electric wheelchair for their specific needs, modifying one chair to operate with cheek rather than hand pressure. With this and other refinements, he made it possible for many World War Two veterans to live richer lives. That alone is a worthy enough deed and legacy. Yet his advance did more: it created the field of rehabilitation engineering, a discipline whose practitioners would help men and women the world over enjoy more mobile and therefore more meaningful lives.

AIDS FOR THE BLIND
The age of independence.

For generations, blind people and those with severely limited sight were often confined to institutions and often lived on the margins. Jim Swail changed that. Blind himself, the Canadian engineer in 1963 led a team of physicists, research scientists, industrial designers, and marketers to innovate hundreds of useful ways for blind people to enjoy more productive, fulfilling lives. These methods — including sensors to detect light sources, sound beacons to locate objects, an improved collapsible white cane, voice synthesizers for phones, a specialized gauge for photo-development vats, and electronic thermometers equipped with readouts that can be heard and felt — improved lives and led to advanced locator technology now used in an array of modern consumer devices. Best of all, Jim created awareness of the needs of the blind and changed public attitudes about blindness on a global scale. His was a case of one man helping all the world see more clearly.

COMPUTERIZED BRAILLE
The palpable library.

Sometimes those who cannot see have the greatest vision. Although Roland Galarneau had a mere 5 per cent of his sight, he envisioned a way for the blind to have much greater access to the written word. In 1972, the electrical engineer from Gatineau, Quebec, designed and assembled a machine he called the Converto-Braille. It scanned and translated text into Braille at a rate of one hundred words per minute. Before Roland built his device, printing in Braille was slow and expensive. Soon after its development, not only did the number of books in Braille jump but Roland's company signed a contract with the education department in Quebec and received a grant to computerize the process. Today, Braille translation software converts texts in many languages and subjects, opening up a new world for the blind. Some people choose to see farther.

CONFEDERATION OF CANADIAN UNIONS

The champion of women's rights.

Madeleine Parent believed unions should promote understanding about the needs of not only workers, but also women — and especially women who worked. A leading figure in Canada's labour movement for some seventy years, Madeleine helped found the Confederation of Canadian Unions in 1969. The group had two primary purposes: to repatriate Canadian unions dominated by American parent organizations, and to advance the rights of women to equality and fairness in and out of the workplace. This second purpose was particularly novel. Unions were then — and had been for generations — dominated by men and the interests unique to them. Madeleine fought to change that, bringing the idea of equal pay for work of equal value to the forefront of the national discussion in Canada and championing the special needs of Aboriginal women and immigrant women. Unions had been around for decades. Madeleine Parent refocused them to serve a new and vital purpose.

RESTORATIVE JUSTICE
The new age of Aboriginal wisdom.

Men and women have spent centuries seeking justice. Two innovators from Elmira, Ontario, believed it required addressing the particular needs of victims, offenders, their families, and their communities rather than simply locking up the guilty. They knew from their First Nations colleagues that there was a better way; traditional restorative justice focused not on punishment but rather on making things right again. So probation officer Mark Yantzi (pictured here) and prison support worker Dave Worth put their belief into action in 1974 when confronted with the case of two youngsters who had vandalized twenty-two properties in their community. Seeking true justice, Mark and Dave asked the court's permission to arrange a meeting between the young offenders and their victims hoping that reparation – amends for wrongs committed – could be made. When the judge agreed, the men wisely turned to local First Nations leaders to better understand the mechanics and lessons of restorative justice practised by countless generations of Aboriginal peoples in Canada. In this case and in many that followed, the modern application of First Nations experience met with outstanding success. Wrongs were righted, wounds were healed, and restorative justice again found favour, paving the way for more formal legal recognition of traditional restorative-justice practices in Aboriginal communities in Canada, the United States, Europe, Australia, and New Zealand. Sometimes, true justice is found first where Aboriginal wisdom meets modern practice.

BLUE BOX RECYCLING
The better way.

Earth-changing ideas are not the property of Ph.Ds. Anyone can innovate if they just look around and ask, "Can't we do this a better way?" Nyle Ludolph asked that question. The Kitchener, Ontario, garbage man was troubled by the vast amounts of waste he saw during his daily pickups, for he knew the landfills in his town were bursting at their seams. His answer came in the form of a simple blue box. In 1983, Nyle championed the world's first municipal curb-side recycling program. His proposed boxes had a distinctive colour to make them visible against either grass or snow, and because blue is most resistant to the sun's damaging ultraviolet rays. His simple and profound idea soon spread across Canada, throughout North America, and around the world, turning what would have been countless tonnes of waste into new products and changing the behaviours and attitudes of millions of people. Turns out there was a better way.

WEEVAC
The stretcher for little people.

Sometimes the simplest question can lead to the most profound results. Wendy Murphy's query arose as she watched television one evening in 1985. On her screen, the Torontonian witnessed rescue personnel in Mexico City trying to transport victims from the rubble of a massive earthquake to the care of nearby clinics and hospitals. Many of these victims were small children, tiny babies even. She asked herself, "Why isn't there an evacuation device designed especially for them?" Taking up the gauntlet she herself threw down, Wendy spent the next two years designing a stretcher for babies and very young children. Made of lightweight aluminum and fire-resistant materials, she called her device WEEVAC. Institutions and jurisdictions around the world soon incorporated the patented stretcher into their emergency evacuation plans. Wendy wasn't done. She went on to found a company that designed and produced other models for a variety of uses and people – especially the littlest among us.

SOLID HONEY
The sweet inspiration.

Sweet mishap. While hiking in British Columbia's bear country in 1990, John Rowe slipped and smashed the jar of honey he was carrying in his backpack. Awkward. But out of one man's misfortune came a honey lovers' delight. The Prince Edward Islander was stirred enough by his fall to develop a solid form of honey that is easily transportable and completely natural, with the look and taste of the original. But he didn't do it alone. After a decade of tinkering, John returned to his home province and enlisted the help of the PEI Food Technology Centre, a hub of expertise related to lab work, testing, business modelling, and marketing. The centre allowed John to perfect a process that removes all water from the honey without changing the nature of the food itself. Together they conducted commercial trials until they knew the new product and business would be a success. The collaboration worked. John's company—Island Abbey Foods—was soon manufacturing hundreds of natural honey products, shipping them to more than twenty countries around the world. The bee's knees.

HOMELESS HUB
The research clearinghouse to end homelessness.

Housing is not a commodity—it's a human right. This principle animates the work of Stephen Gaetz. The York University professor of education leads the Canadian Observatory on Homelessness and is founder of the Homeless Hub, the world's first comprehensive, cross-disciplinary, web-based clearinghouse of research on homelessness. It's used by governments in Canada and around the world to develop policies and practices that not merely manage homelessness but also prevent, reduce, and end it for good—because each of us has the right to a home.

IMPROVE A PRODUCT: HERE'S HOW.

☐ Find a product or object that needs improving

☐ Identify features of competitive products, and make a list of desirable improvements.

☐ Be honest about which improvements your customers need and would pay for.

☐ Figure out what resources of money, people, and time you need to make the top improvements. Start working only when you have those resources in place.

☐ Work through many prototypes, noting how and why things fail, sharing your findings with your whole team.

ARGAN OIL COOPERATIVE
The resourceful enterprise.

Women are the world's greatest untapped resource for peaceful, prosperous, sustainable development. This knowledge is the inspiration behind the Argan Oil Cooperative. Developed in 1997 by the International Development Research Centre of Canada, the group is made up of more than a hundred Moroccan women who harvest and process oil from the kernels of local argan trees — which for years had been another under-used resource — and then sell the edible oil to consumers around the world. Proceeds from sales enable the women to support their families and strengthen their communities. The women also work to preserve the long-term health of the trees. The argan not only supplies the women with their livelihoods but also serves as a bulwark against growing desertification, which is the greatest natural threat to peaceful, prosperous, sustainable development in the region.

BREAKFAST FOR LEARNING
The nutrition of academia.

Nutritious food fuels not only the body but also the mind. Children who enjoy a good breakfast are ready to tackle the learning challenges of their school day. Children who arrive at school hungry are able to think about one thing only — how hungry they are. First set up in Toronto in 1992, Breakfast for Learning was the world's first national program to help schools make sure kids get the nutritious breakfasts and lunches they need. Since its start, the program has been a smashing success, partnering with some 1,600 schools across Canada to serve nearly 600 million meals to just short of four million young students. Breakfast for Learning is showing the world that schools have roles to play not only to instruct but also to equip students to learn. It's teaching the world that the path to learning runs through a full stomach.

CRADLEBOARD
The border crossing.

"Who am I and who do others think I am?"
That question lies at the heart of Cradleboard.
Developed by teacher and entertainer Buffy
Sainte-Marie in 1996, Cradleboard is a series of
three curricula for children in elementary,
middle, and high school made up of courses
about music, history, science, geography, and
social studies. All three learning units give
students — who can be non-Aboriginal as well as
Aboriginal — an Indigenous perspective about
each subject. These perspectives, which are
widely neglected in conventional curricula,
promote self-esteem in Aboriginal learners and
spark empathy and understanding among
non-Aboriginal students. It is no mystery that
Buffy Sainte-Marie is the mind behind
Cradleboard. An internationally renowned
performer, she knows first-hand how the act of
crossing borders reveals new perspectives and
leads to greater understanding.

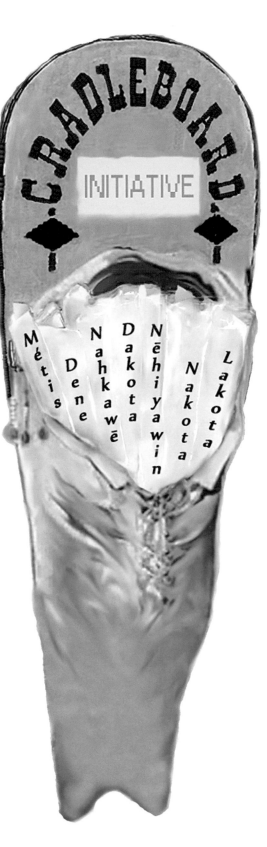

NUNAVUT
The new way to govern.

How can a country recognize the historic rights of an Indigenous people to their heritage lands and waters while staying committed to pluralist government in which all citizens have an equal voice? One Canadian answer was Nunavut. The word means *our land* in Inuktitut. This third and largest territory of Canada was created in 1999 by an act of Canada's Parliament and an agreement between the country's government and representatives of the Inuit people. The act of Parliament defined a territory with its own democratic government within Canada's federal system, while the agreement superimposed that government over the largest Aboriginal land claim in the world. The creation of Nunavut, just like the creation of Canada itself, was an innovation intended to uphold the rights and fulfill the hopes of all. Both were new ways of thinking about governing a land.

ABEEGO
The beeswax wrap.

Change the way people store food. How's that
for an ambition? It's the one behind Abeego, and
Calgary's Toni Desrosiers is the nutritionist and
entrepreneur who conceived it. In 2008, she
created the all-natural wrap — a combination of
hemp and cotton fabric infused with beeswax
and tree resin — to be the most sustainable way
to preserve food. Why Abeego? People have been
storing food for as long as we've had people.
Today's people use plastic wrap. It can be tricky
to manipulate, doesn't always do the job, and,
worst of all, generates a tremendous amount of
non-biodegradable waste. Abeego is the antith-
esis of all three — it's naturally adhesive; it's
antibacterial, antimicrobial, and breathable so it
keeps food fresh; and it's completely reusable
and biodegradable. It's the first wholly natural
solution to the artificial problem of plastic wrap.

ME TO WE
The first social business.

Can hard-headed business principles be used to
achieve kind-hearted international-development
goals? The answer is yes, and Me to We is proof.
Me to We began in 1999 as a for-profit enterprise
that ran camps to nurture emerging leaders and
organized volunteer trips to developing countries. Proceeds from these activities were used by
Free the Children — a Canada-based international-
development organization — to help young
people and their families around the world enjoy
safer, healthier, happier lives. Both Free the
Children and Me to We are the brainchild of
Craig and Marc Kielburger — two Canadians who
have devoted their lives to serving others. Today,
the world's first social enterprise has grown
from supplying camps and trips to offering a
range of consumer goods, educational resources,
and learning experiences, with all money
generated still going to Free the Children. The
work done by Me to We is celebrated each year
through We Day. What began as a single event is
now a series held in cities across Canada, the
United States, and the United Kingdom. We Day
brings together a total of some 200,000 young
volunteers to celebrate their achievements,
inspire themselves and others, and show that
business principles can be used to create social
enterprises and make a world of difference.

SAKKIJÂNGINNATUK NUNALIK
The truly northern home.

Housing is the most pressing need for nearly all northern communities in Canada. The problem stems from a lack of safe, healthy, energy-efficient homes designed and built by Inuit for Inuit. That's changing. SakKijânginnatuk Nunalik, which is Inuktitut for *sustainable communities*, is using thousands of years of Inuit knowledge and the best of modern practices to create the world's first sustainable, multi-unit residential dwelling for the north. By combining time-honoured wisdom and the best current methods, the structures will satisfy the preferences of Indigenous residents, account for the region's changing climate, and address a most pressing need.

MILK CARTON 2.0
The better way to find children.

The milk carton campaign to locate missing children worked for previous generations because it centred on something that nearly all people held in their hands at some point during each day — a carton of milk. Today, we hold something else in our hands not merely once or twice each day but almost continually — a smartphone. The Missing Children Society of Canada's Search Program is the world's first social digital program to find missing children. Created in 2014, the application (also known as Milk Carton 2.0) engages people, corporations, and the media via its three components: Most Valuable Network uses the viral power of social media to alert the public to a missing child; CodeSearch takes advantage of geo-targeted

alerts and real-time news feeds to engage corporate partners and their employees with law enforcement officials in the active search for missing children; and Marketwired pushes notifications to thousands of traditional and digital media outlets across the country. Generations ago, milk cartons were the better way to locate missing children. Today, our milk carton is digital, social, and mobile.

RIGHT TO PLAY
The natural classroom.

Play is more than mere fun. It's an instinctive and essential way for children to grow physically, cognitively, and emotionally. It can also be used to teach kids in developing countries specific lessons they can use to improve their lives. That's the ethos of Right To Play, an international-development organization founded in 2000 in Toronto by Johann Olav Koss. Right To Play's teaching method takes the childhood instinct to play and channels it so youngsters reflect on the experience of the game they just finished enjoying, connect what they experienced during it to a similar experience from their own lives, and explore how they can apply what they've learned to an area of their lives. In so doing, Right To Play's curriculum teaches children life skills that can help them overcome the effects of poverty, conflict, and disease. The organization's method is now at work in developing countries around the world, enabling more than one million kids to experience the life-affirming lessons, joy, and right to play.

TRUTH WITH RECONCILIATION
The healing touch.

The ability to turn something sorrowful into something hopeful is one of the most vital forms of human alchemy. The Truth and Reconciliation Commission (TRC) was set up in 2008 to come to grips with the damage done to generations of Indigenous children sent to Canada's residential schools. Conditions in the schools were appalling. The children were subjected to sustained cultural, physical, and spiritual deprivation and injury. Many were sexually abused. Many lost their lives. This has been a hard truth for Canadians to accept about their country. Yet Justice Murray Sinclair and his fellow commissioners, through an extraordinary blend of clarity, patience, compassion, and good humour, were somehow able to capture the imagination of Canadians with a vision of reconciliation and renewed partnership. Over a six-year period, they spoke with more than 9,000 survivors

through 238 days of local hearings in 77 communities across Canada, and at numerous national events. The exercise was physically and psychologically exhausting for everyone concerned, yet it resulted in a vast array of recommendations to address the long-standing imbalance between the people of Canada in general and the First Nations people within that population. Commissioner Justice Sinclair is shown here, embracing a residential school survivor after she spoke at the release of the final report of the Truth and Reconciliation Commission in 2015. Standing on the shoulders of the Royal Commission on Aboriginal Peoples, the TRC managed in 2015 to finesse what will likely prove a dramatic turning point in the country's capacity to tap the ingenuity made possible by its Indigenous inheritance.

ART AS INNOVATION
The living memorial.

Art can be a powerful reminder of loss and an empowering assertion of presence. Christi Belcourt's work as an artist, an activist and a catalyst for Indigenous self-determination make her one of the most formidable Indigenous voices in the country. Her artwork is widely collected and admired for its beauty and its depiction of Indigenous worldviews. Compelling and rich, it is a window into Indigenous connection to land, to tradition and to all life. Her community based engagement, as well as her work with the Onaman Collective poignantly and powerfully mobilizes Indigenous resurgence, claiming, reclaiming and proclaiming Indigenous presence and place on this land. Christi was one of the first recipients of the Governor General's Innovation Awards, presented in 2016.

LEAD A SOCIAL CHANGE: HERE'S HOW.

☐ Form a team of change leaders from all sectors and levels of the community or organization.

☐ Create a clear vision of how things will look when the change is made, and establish a sense of urgency.

☐ Empower others to take action that helps achieve the vision, but let them choose how. Let them communicate the vision far and wide.

☐ Define, plan for, and measure short-term success.

☐ Consolidate the improvements and create still more change.

☐ Imbed the new approaches into the community culture by celebrating and repeating.

* Adapted from *Leading Change* by John P. Kotter, 1990

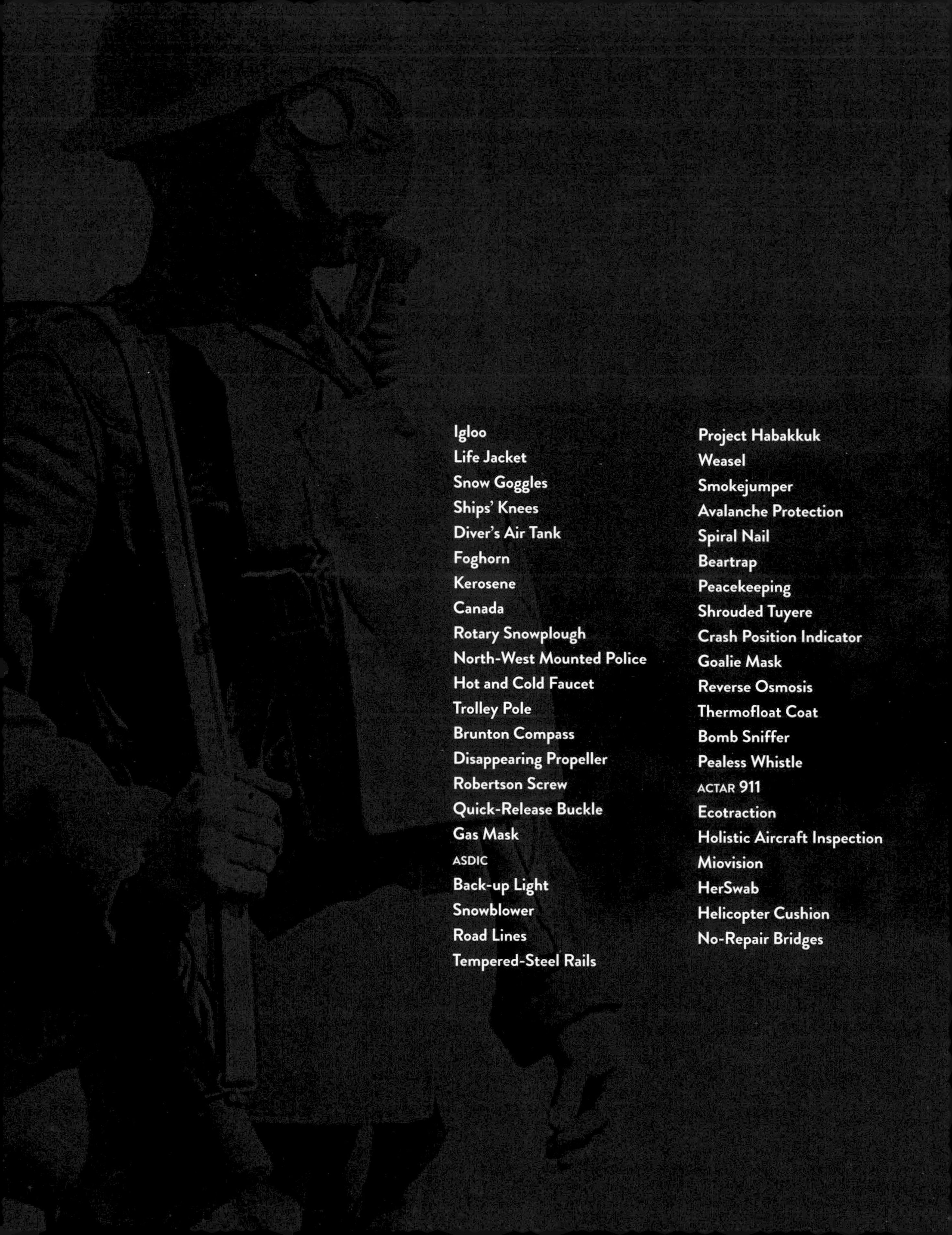

Igloo

Life Jacket

Snow Goggles

Ships' Knees

Diver's Air Tank

Foghorn

Kerosene

Canada

Rotary Snowplough

North-West Mounted Police

Hot and Cold Faucet

Trolley Pole

Brunton Compass

Disappearing Propeller

Robertson Screw

Quick-Release Buckle

Gas Mask

ASDIC

Back-up Light

Snowblower

Road Lines

Tempered-Steel Rails

Project Habakkuk

Weasel

Smokejumper

Avalanche Protection

Spiral Nail

Beartrap

Peacekeeping

Shrouded Tuyere

Crash Position Indicator

Goalie Mask

Reverse Osmosis

Thermofloat Coat

Bomb Sniffer

Pealess Whistle

ACTAR 911

Ecotraction

Holistic Aircraft Inspection

Miovision

HerSwab

Helicopter Cushion

No-Repair Bridges

FOUR

Smarter
Smaller
Kinder
Safer
Healthier
Wealthier
Happier

Life is a risky business. Everywhere we live and every way we interact poses daily threats to our comfort, health, and, ultimately, our lives. Where safety is concerned, Canadians have collaborated in novel ways to reduce the risk in everyday situations. Their contributions have given the world effective means to avoid drowning, burning, freezing, crashing, asphyxiating, skidding, colliding, slipping, running aground, and being poisoned, car bombed, or shot.
Thanks, eh.

IGLOO
The ingenious structure.

Engineers and architects have found few structures more intriguing than the traditional home of the Inuit peoples of Canada. As old as human civilization itself, the igloo is ideal for several reasons. Perfect in composition, an igloo is made up entirely of the substance most abundant in Canada's North – snow. Hard-packed snow cut into bricks – three feet by two feet by four inches thick – form the igloo's structure. After the first row of snow blocks has been placed, several consecutive blocks are cut to form a sloping surface. A second row is then placed on the first, and then the blocks are built up in a continuous spiral until the final one is placed at the very summit. Snow also acts as mortar to seal any cracks and prevent cold air from entering and warm air from escaping. Perfect in design, an igloo has a rounded beehive shape – known as a catenoid – that has no corners and therefore no wasted space. Nearly circular with a slight inward lean that becomes more acute as the wall rises, this shape enables the structure not only to bear enormous amounts of weight but also to become stronger as more weight is applied to it. Perfect in suitability, an igloo can be a permanent or temporary structure for one or many occupants. Those inside the structure warm it with the heat that emanates from their bodies. A small hole in the roof serves to ventilate exhaled breath from the people inside. Another small hole near the door allows fresh air to enter. Perfect in efficiency, an igloo is put together by two people in under an hour. The only tool required for the job is a long knife of bone or ivory. No wonder the igloo has lasted for millennia.

LIFE JACKET
The Inuit fisher's insurance.

When exposed to Canada's frigid waters – both coastal and inland – you will often perish more quickly from heat loss than drowning. Inuit whale fishers knew this truth. They made what are known as spring-pelts, which are sealskin or seal gut stitched together to create a waterproof covering for their torsos. These early life jackets evolved, more insulated and buoyant over time, until they became the sailor's salvation we know today.

SNOW GOGGLES
The answer from the world around them.

Anyone who has walked across a snow-covered field on a sunny day can attest to the blinding intensity of the glare. Today, we quickly slip on our Ray-Bans or Oakleys. The Inuit people of Canada's North had no such luxury. For an answer to snow blindness, they looked to the world around them. Taking a piece of antler or bone, usually from a caribou, the Inuit carved a long slit through it—wide enough to give its wearer the ability to see but narrow enough to block most of the harmful and harsh ultraviolet rays. For the goggles to work reliably, they needed to allow in only the light coming through that slit. So the Inuit curved them to fit the contours of the wearer's face, carved out a piece at just the right spot to accommodate the nose, and fastened them securely behind the head with a cord made out of caribou sinew—another answer from the world around them.

SHIPS' KNEES
The mariner's insurance.

Isolation often leads to innovation. Thousands of kilometres from their homeland, settlers in New France built their own ships. In 1748, when shipbuilders in the port of Quebec were still using curved pieces or *knees* of cut wood to secure the hull beams of ships at odd angles, one blacksmith had an idea. Why not improve these hard-working ship's knees by making them at a variety of angles out of iron? Working through a series of welded prototypes, the blacksmith's team developed an early quality-control test. If a welded knee could be dropped from a height of fifty feet onto the barrel of an iron cannon without breaking, it was deemed ready for sea duty. Although the early innovator's name is lost to history, his advance was so effective it became a regular shipbuilding practice far beyond the isolated wilds and waters of New France and is credited with transforming the shipbuilding industry throughout North America.

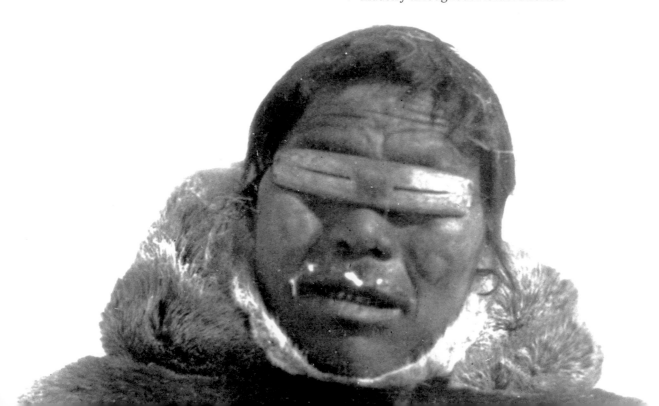

DIVER'S AIR TANK
The precursor to scuba.

They called it an oxygen reservoir for divers. James Elliott and Alexander McAvity—two men from Saint John, New Brunswick—created in 1839 what we would recognize as the first scuba tank. Their equipment was made up of a copper vessel that carried a quantity of condensed oxygen, a pipe from the vessel to the interior of a diver's suit, and a stopcock to regulate the flow of the oxygen from vessel to interior. Tank, tube, and regulator: the precursor to scuba.

FOGHORN
The coded voice of danger.

The lower the frequency of a musical note, the farther that note can be heard. Simple physics. Robert Foulis grasped this fact for himself one foggy evening in his home of Saint John, New Brunswick, in 1853. Walking toward his house, he heard his daughter playing the piano. As the notes drifted on the night air, he noticed he could hear the lowest notes clearest. Well aware of the dangers of fog to shipping, Robert put his personal discovery to good use. Knowing that lighthouses can't warn sailors when their beams cannot be seen, he devised the steam foghorn— a megaphone designed to blast a coded set of low-frequency notes to warn mariners of particular dangers close to shore. Foulis shared his innovation with provincial lawmakers. Not long after, the first foghorn was installed near his home at the Partridge Island lighthouse, from which its coded voice of danger would be heard for nearly 150 years. Although Foulis never patented nor profited from his innovation, fogbound mariners to this day keep their ears cocked for the welcome sound of his life-saving idea.

KEROSENE
The safe illuminating gas.

For generations, coal gas lighted the dark. Coal was plentiful and relatively easy to get at, but the gas it produced was dirty and dangerous. It burned with a smoky flame and could combust easily. At his Parrsboro practice, the Nova Scotian physician Abraham Gesner had treated horrific burns on parents and children whose homes had been set ablaze by coal-gas fires. Gesner was also a geologist and combined his two disciplines to seek a solution. In 1846, he gave a public demonstration of an exciting innovation. He heated coal and distilled from it a clear, thin liquid that could fuel lamps. He called the liquid kerosene. When lit, kerosene burned much cleaner than coal gas. It was also much more stable and therefore safer, not prone to combust like its gaseous cousin. Eight years later, Gesner patented the kerosene process and the safe illuminating gas was on its way to lighting the nighttime world. Characteristically, he took great pains to redirect all praise, saying, "The progress of discovery in this case, as in others, has been slow and gradual. It has been carried on by the labours, not of one mind, but of many, so as to render it difficult to discover to whom the greatest credit is due." Kerosene was only one of Gesner's many original ideas. While a resident of Saint John in neighbouring New Brunswick, he opened Gesner's Museum of Natural History, Canada's first public museum. We know it today as the New Brunswick Museum.

BUILD AN INNOVATION ECOSYSTEM: HERE'S HOW.

☐ Connect with organizations and people who believe that innovation is necessary and good.

☐ Find partners who share your values, and who will prosper through their relationship with you.

☐ Build groups of partners within your particular region (e.g., Waterloo, Ontario), platform (e.g., iOS), or industry (e.g., fast food).

☐ Find suppliers who can ensure a steady flow of parts and materials at competitive prices, even when you scale up.

☐ Test your ecosystem with a small project that involves each partner, and note any functions that you're missing or duplicating.

☐ Don't get greedy. Make sure that successes in all your joint projects benefit *all* players in the ecosystem.

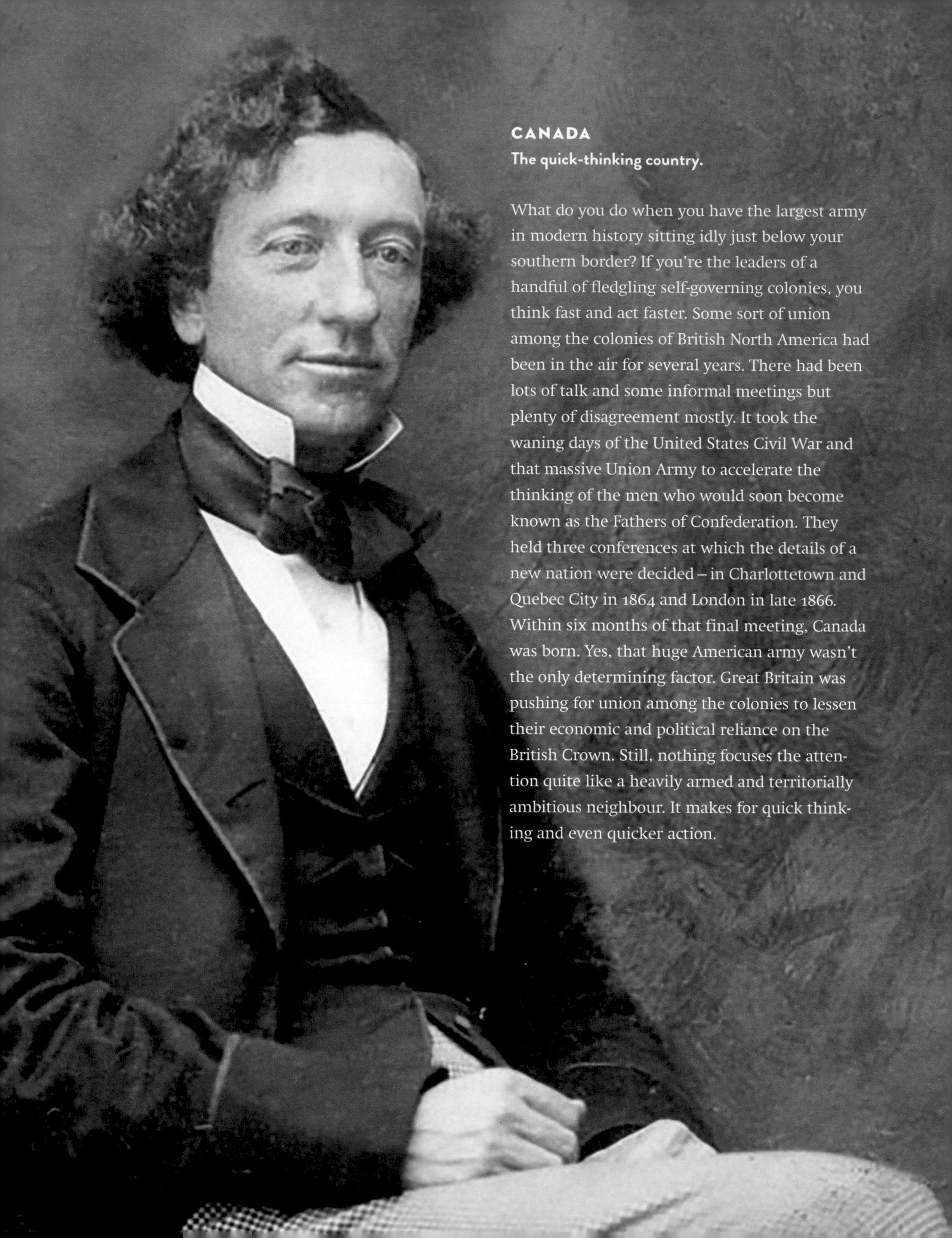

CANADA
The quick-thinking country.

What do you do when you have the largest army in modern history sitting idly just below your southern border? If you're the leaders of a handful of fledgling self-governing colonies, you think fast and act faster. Some sort of union among the colonies of British North America had been in the air for several years. There had been lots of talk and some informal meetings but plenty of disagreement mostly. It took the waning days of the United States Civil War and that massive Union Army to accelerate the thinking of the men who would soon become known as the Fathers of Confederation. They held three conferences at which the details of a new nation were decided — in Charlottetown and Quebec City in 1864 and London in late 1866. Within six months of that final meeting, Canada was born. Yes, that huge American army wasn't the only determining factor. Great Britain was pushing for union among the colonies to lessen their economic and political reliance on the British Crown. Still, nothing focuses the attention quite like a heavily armed and territorially ambitious neighbour. It makes for quick thinking and even quicker action.

ROTARY SNOWPLOUGH
The winter wonder.

Is anyone surprised that the first railway snow-plough was conceived in Canada? In 1870, J.W. Elliott created a device to clear snow from railway tracks. The Ontarian's plough used a steel collector that fed snow into fan plates on the edge of a wheel. The wheel, driven by a rotary engine, threw the snow out the top of the wheel casing far away from the tracks. Another Ontario innovator, Orange Jull, improved Elliott's device by placing a cutting blade in front of the fan plates. Mounted on the same shaft and spinning in the opposite direction, the blade cut up the snow and made the work of the fan easier. Within twenty years, the rotary snowplough was in widespread use on Canadian rail lines. By 1911, approval had been given for a plough with a twelve-ton blade that could cut through any volume of snow, and also any rocks and trees mixed in. A version of this plough is still used by railways in Canada and the United States.

NORTH-WEST MOUNTED POLICE
The mild, wild west.

Many frontiers have been breeding grounds for anarchy — the violence that can quickly erupt when no authority is present. Not so Canada's Northwest. Credit for the calm goes to the North-West Mounted Police. Just a few years after Confederation, Prime Minister John A. Macdonald took the recommendation of military advisers that a new force be established as an innovative blend of cavalry regiment and constabulary. With protection of peaceful society as its ultimate aim, the North-West Mounted had a delicate balance to keep: preserve order, assert Canadian sovereignty over the vast Northwest Territories, enforce the existing agreements with the region's First Nations, and stop American whisky traders. Critical to the mandate, these duties had to be carried out without any excessive force. The new outfit first mounted up in Fort Dufferin, Manitoba, in 1874 as a unit of 22 officers and 287 men and boys aged fourteen and up. The force proved able and fair and was soon widely respected. Their presence in the region had a host of benefits, including forging peaceful relations between the Crown and First Nations and allowing the orderly development of settlements in communities along the proposed route of the transcontinental railway. With a distinctive force clad in red serge and not the gun: that's how the Canadian West was won.

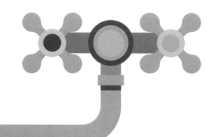

HOT AND COLD FAUCET
The source of warm water.

So many aspects of modern life are taken for granted. Flip a switch and light fills a room. Adjust a dial and red-hot coils cook a meal. Lift a tap and warm water flows uninterrupted. This last one has a surprisingly long life. The first combined hot and cold faucet was created in 1880. Thomas Campbell of Saint John, New Brunswick, is the man behind the advance. A fundamental challenge he had to solve in bringing his faucet to life was overcoming back pressure. If the faucet failed to discharge the hot and cold water quickly enough, back pressure at the spout would prevent more than a trickle of water from emerging. Thomas overcame this problem by making sure the single spout had a larger capacity than the amount of water discharged by cold and hot valves combined. Problem solved. The source of warm water had arrived.

TROLLEY POLE
The other heavenly power.

Electric street railways were a great way to get around. But only on sunny days. The electricity that powered these fair-weather friends was fed to their motors through a third rail buried under the pavement. Heavy rains would ground the underground wire and the cars would grind to a halt. In 1883, James J. Wright, the operator of a Toronto street railway, took the power from underneath the cars and raised it above them. The device that made it all possible was Wright's trolley pole. The pole, attached to the top of the car, took electricity from overhead wires and brought it to the cars. Each pole was equipped with a wheel to reduce friction between pole and wire as the car rumbled down the rails. He demonstrated his system at the Canadian National Exhibition that summer. For two weeks the cars ran — in sunshine and rain — without once breaking down. Soon the overhead wires and trolley poles were fixtures on streetcars in cities wherever the sun shone and the rain fell.

BRUNTON COMPASS
The new sense of direction.

Innovation often involves taking an indispensable piece of equipment and making it smaller and simpler while losing none of the wholeness of the original. David Brunton practised this elemental action of innovation. In 1894, the Canadian-born geologist and mining engineer created the Brunton Compass. Also known as the Brunton Pocket Transit or simply the Brunton, the precision compass is a compact version of the heavy surveying equipment that engineers had to haul into the field to navigate and take accurate measurements. The Brunton, which uses magnetic induction rather than fluid to damp needle oscillation, remains the most popular compass of geologists. It's also relied on by surveyors, engineers, archaeologists, and just plain outdoorsmen — pretty much anyone looking to go in the right direction.

DISAPPEARING PROPELLER
The boater's advantage.

Canada is surrounded by three oceans and enjoys the second longest coastline in the world. Yet what is perhaps more impressive is the number and variety of lakes and rivers in the country. The retreating glaciers of the Ice Age left behind a boaters' paradise. Harold Wilson made this weekenders' Eden an even more tempting place in 1900 when he developed the disappearing propeller. Using a lever located near the steering wheel, the captain of a small boat could lift the propeller and propeller shaft into a well in the boat's hull. A boat equipped with the new prop would now be able to negotiate waters with shoals and hidden rocks so common in the lakes near Harold's Gravenhurst, Ontario, home. No more dented props. No more inaccessible fishing spots. Weekend boaters had found their advantage.

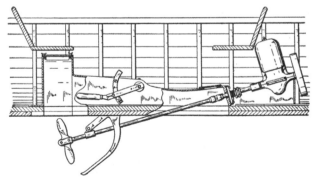

The Perfect Convertible Boat at Last.

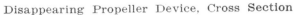

Disappearing Propeller Device, Cross Section

ROBERTSON SCREW
The no-slip screw and driver.

Peter Robertson cut his hand and changed an industry. The mishap occurred in 1908 in Montreal, when the Milton, Ontario, salesman for a Philadelphia tool company tried to demonstrate how to use a spring-loaded screwdriver to fasten slotted screws. Slotted screws—more commonly known by the surname of their creator, Henry Phillips—were notorious for not only causing screwdrivers to slip but also for being stripped themselves when fastened or unfastened. Once his hand healed, Peter got to work. The industry-changing device he came up with is a square-socket screw. The screw's chamfered edges, tapering sides, and pyramidal bottom meant it could be screwed in faster, easier, and tighter than Phillips's version. The eponymous Robertson was an immediate hit. Henry Ford, for one, insisted on using Robertson screws and screwdrivers when he learned that his automotive assembly lines could build each car two hours faster doing nothing different but using these new screws. Everyone wanted a piece of the action, but Peter wasn't willing to hand over control. As a result, the Robertson screw never became the universal phenomenon it deserved to be. That said, more than a century later the company that bears his name is still producing his superior patented screws for everyone who wants a safer, easier, tighter fit.

QUICK-RELEASE BUCKLE
The fast way into the saddle.

Most of the technologies of the horse and buggy were swept aside with the arrival and embrace of the automobile. The quick-release buckle remains. In fact, it grows more popular with time. Arthur Davy of New Glasgow, Nova Scotia, created the device in 1911 as a way to attach the ends of two reins quickly and securely—and to detach them equally quickly and smoothly. Instead of the conventional buckle, Davy's version featured one part equipped with a pair of spring-actuated dogs (a dog is a mechanical device for gripping) adapted to extend into and engage slots in the second part. Click! That sound is heard now more than ever, as people fasten tight clothes, car seats, briefcases, backpacks—anything that requires a quick, secure click shut.

GAS MASK
A new helmet for Hell.

Out of horror springs ingenuity. World War One produced more than its fair share of horrors. Poison gas stands at the head of the list. The German army used the gas for the first time at Ypres, Belgium, in 1915 against Canadian and French colonial troops. Dr. Cluny MacPherson, acting quickly out of obvious necessity, created a mask to defend Allied troops against further attacks that used the deadly gas. His creation was an amalgam of materials at hand: a helmet taken from a captured German fitted with a makeshift canvas hood that featured eyepieces and a breathing tube. The entire apparatus was treated with chemicals to absorb chlorine, the lethal element of the gas. The device created by the principal medical officer of the St. John Ambulance Brigade of the Royal Newfoundland Regiment turned out to be the most protective of the entire four-year cataclysm. Shielding countless men from death, disfigurement, or permanent illness, it was one Canadian's saving grace in the midst of Hell.

Newfoundland became part of Canada in 1949.

ASDIC
The sub hunter.

Call it the supreme irony of innovation. A Canadian was responsible for one of the most important technological advances of World War One, yet his name and role in creating what we know today as SONAR are virtually lost to history. As part of the British effort to combat German U-boats, physicist Robert William Boyle in 1917 combined transducers made with quartz (a material particularly effective for the radiation of high-frequency sound) with amplifiers developed by French researchers to create the world's first high-frequency echolocation device. The device known as SONAR (SOund Navigation And Ranging) sends sound waves into water; once the waves strike an object, they echo back. The distance of the object is measured by calculating the amount of time between when the signal is sent and the echo received. SONAR can also determine the range, bearing, and speed of an underwater vessel. Much of the credit, if we can call it that, for Mr. Boyle's descent into this historical black hole has to do with wartime secrecy. The name of the project — ASDIC (shorthand for Anti-Submarine Division) — was deliberately meant to camouflage any mention of sound or quartz. And Newfoundland-born Boyle himself never took out patents for his creation, nor published any papers, nor received any formal recognition. The sub hunter operated in stealth.

BACK-UP LIGHT
The safer rear view.

James Ross had a bright idea. Literally. In 1919, the Halifax automobile enthusiast affixed a light to the rear of his vehicle and rigged it to a switch at the base of the gearshift. When he shifted his car into reverse, the light went on. It made driving at night safer for him and all those around him. Still does. His idea lights up our nighttime roads – brilliantly.

SNOWBLOWER
The return to the open road.

Of all the Canadian efforts to overcome the restrictions of cold weather, the innovation of Arthur Sicard of Saint-Léonard-de-Port-Maurice, Quebec, may have made the most difference. Sicard hatched an idea back in 1894 when just eighteen years old, but it wasn't till he was almost fifty that he found the time to produce a prototype. He called it *la déneigeuse et souffleuse à neige Sicard*, or the Sicard Snow Remover Snowblower. The device combined a four-wheel-drive truck chassis, truck motor, snow scooper, and blower with two adjustable chutes and separate motor. It was the first commercial-grade snow-removal device in the world. The machine threw snow almost a hundred feet from the truck, or directly into the back of the truck in tight situations. The snowblower was an instant hit; by 1927 his vehicles were removing snow from the roadways of the town of Outremont, a suburb of Montreal.

ROAD LINES
The end of highway chaos.

You would think dividing lines on roads have
been around for as long as paved roads themselves. Not so. It wasn't until 1930 that the first
dividing line was painted on a road. The stretch
of asphalt in question was a portion of highway
near the border between Ontario and Quebec.
The man behind the white stripe was John
Millar, an engineer with Ontario's department of
transport. Not surprisingly, the innovation
caught on quickly and soon dividing lines were
standard features of roads across North America.
Over time, dashes, yellow lines, double lines, and
other variations were added as ways not merely
to divide traffic but also to communicate with
drivers. Talking road lines – an innovation on an
innovation.

PROJECT HABAKKUK
The airstrip in the sea.

Not all attempts to innovate succeed, yet sometimes their foresight and ambition are worth celebrating. Project Habakkuk is one of those noble failures. Prompted by massive losses of Allied supply ships to German submarines, the idea – conjured up by British Combined Operations Headquarters in 1940 – was to carve floating airfields out of thick slabs of ice. Sub-hunting aircraft would use the icy airstrips to protect supply ships traversing the North Atlantic. Fighters and bombers would also take advantage of the frozen airfields to attack parts of Europe. The airfields themselves, bobbing about in the cold Atlantic waters, would be difficult for enemy submarines to spot, could be easily repaired, and could be moved about according to changing wartime needs. A team of Canadian engineers and scientists at Patricia Lake in northern Alberta began construction on a massive model of the proposed airstrip. When they found the ice split too easily, they strengthened it with pykrete – a new material made of wood pulp. Project Habakkuk was halted, however, when other, more realistic means of protecting supply ships emerged, including radar, convoy systems, and longer-range aircraft. Yet all did not melt away. The project led researchers to gain new knowledge about the properties of ice, and pykrete to this day is a strengthening component of remote northern runways. Project Habakkuk: the noble failure that wasn't.

TEMPERED-STEEL RAILS
The safer track.

Slow down the cool-down. In 1932, Irwin Mackie of Dominion Steel and Coal Corporation in Sydney, Nova Scotia, developed a process to stop shatter cracks from appearing in the steel rails of train tracks. The Mackie cooling method involves slowing the rate at which newly milled rails cool by making sure the temperature of the cooling rails stays within certain ranges at certain times. Within ten years, most of the world's rail producers were using the Sydney metallurgist's method. The slow-down cool-down didn't take long to heat up.

WEASEL
The versatile weapon.

Leave it to a Canadian to design the ultimate military snow vehicle. The Weasel is its name. The assault vehicle was created in 1942 to support what was to be an Allied invasion of Nazi-occupied Norway. Planners brought master inventor George Klein on board to design the vehicle's tracks. They picked the right man. Working at Canada's National Research Council, George was an accomplished mechanical engineer who had done extensive research on the properties of snow and ice. When he recommended the vehicle be equipped with a light, all-rubber track flexible enough to shed snow and ice, manufacturers at Studebaker in Indiana balked. They preferred metal tracks. Bad move: the metal tracks iced up during tests. George's design won out, but the Weasel never made it to Norway. While that invasion was called off, more than fifteen thousand of the vehicles did see action in other theatres and remained popular after the war in the Arctic, Antarctic, and wherever else the snow flies.

SMOKEJUMPER
The airborne firefighter.

The best way to fight a huge forest fire is to make sure it doesn't get huge in the first place. In 1947, Saskatchewan's department of natural resources took this idea and combined it with the wartime advent of paratroopers to create the world's first smokejumpers. The name is literal: these parachuting firefighters drop on a blaze while it's still small enough to be put out with saws, shovels, axes, and hoes. It's a job for the stout-hearted, and the original smokejumpers were young, physically fit, and trained in first aid, fire control, parachute packing, and, of course, parachuting. They were a paramilitary force whose enemy was fire. Like the conflagrations these men were committed to stop, the idea they embodied spread beyond Saskatchewan far and fast. Smokejumpers flared up in the northwest United States and eastern and central Russia — regions thick with forests but sparse in roads. Smokejumpers — ironically enough — are a small thing that became huge.

TURN YOUR IDEA INTO A BUSINESS: HERE'S HOW.

☐ Develop your idea into a fully formed concept, then simplify it so you can easily describe it to others.

☐ Discuss your idea with people you trust, then identify flaws and make improvements.

☐ Incorporate a company.

☐ Build a prototype, often called a minimal viable product (MVP).

☐ Discover what your customers' problem is and find out how many customers have that problem.

☐ Spell out the value that solving the problem will create for those customers.

☐ Refine your prototype until you have a solid product that directly addresses the problem.

☐ Have your customer use the product (even in prototype form) to see if it really does solve the problem.

☐ Ask your customers if they will pay for the minimal viable product. If not, improve it again and again until they say they want it. When customers will actually pay for your product, you have a business.

AVALANCHE PROTECTION
The life-saving collaboration.

What does it take to cheat white death? Teamwork. Starting in 1950, highway builders, transportation officials, and snow-mass researchers brought their special knowledge and talents to bear to prevent avalanches from claiming the lives of motorists who travel the Trans-Canada Highway through Rogers Pass. Located in British Columbia, the pass is a notoriously dangerous spot for avalanches. One in 1910 killed sixty-two railway workers. As the new highway was being planned, avalanche-control experts at Canada's National Research Council helped builders select the safest route through the pass. They also worked with officials, teaching them how to forecast avalanches by observing wind, temperature, and snow cover. Today, they are still on the job, working with locals to maintain the world's largest mobile avalanche-control program. And what of white death? No motorist has succumbed to it at Rogers Pass since this stretch of the highway was completed in 1962.

SPIRAL NAIL
The Ardox.

We all have our obsessions. Allan Dove's was nails. Growing up, he would pound them in and pull them out. All kinds of them: old ones and new, tiny tacks and square spikes, round nails and cut nails. When he reached adulthood, he started to study nails more closely and think about ways to improve them. He reached the apex of his obsession in 1954 when he created the spiral nail. As an employee of Stelco — Hamilton, Ontario's steelmaking giant — he realized he and his colleagues would be better served by a nail that was easier to hammer than a smooth, round nail; one that didn't split wood when pounded in; one that gripped more strongly once it was in. His spiral nail did all three: it went in fast, rarely split wood, and gripped tight. With an obsession, some might say.

BEARTRAP
The sure way down.

Some beartraps are about as far away from the forest as you can get. The beartrap used by navies around the world — formally known as the Helicopter Hauldown and Rapid Securing Device — is used to land helicopters on the rolling, pitching flight decks of frigates, destroyers, and other smaller warships. Developed in 1956 by Fairey Aviation in Dartmouth, Nova Scotia, the beartrap may be located on a ship, but it is really a conduit to bring ship and helicopter together. Picture a helicopter hovering over a moving deck. It lowers a line with a probe on the end. The deck crew attaches the probe to a heavier cable, which has been passed through the beartrap from a winch below deck. The combined line and cable are pulled up until the cable is secured to the helicopter. The pilot then increases power to balance the pull of the winch with the lift of the helicopter. This move puts the aircraft in synch with the movements of the ship. The pilot then decreases power and the winch slowly and surely pulls the helicopter to a position just above the deck. Rough seas may be pitching the helicopter and ship about, but the two are now doing their dance in harmony, swaying together. When the moment is right, a ship's officer instructs the pilot to land. The beartrap then closes, capturing the helicopter's main probe, securing the aircraft to the deck. The device isn't finished its work, though. If need be, the beartrap then tows the helicopter into its hangar. Navies in Japan, Australia, the United States, and elsewhere were quick to latch on to the beartrap. As Canada's greatest contribution to naval aviation, it became for them as well as Canada the sure way down.

PEACEKEEPING
The man who saved the world.

Soldiers fight wars. It took one mild-mannered Canadian to see that they could also be instruments of peace — not a peace of victory for one side and defeat for the other, but a peace brought about by both sides having the confidence and assurance to cease fighting. The man was Lester Pearson. The Canadian diplomat stepped onto the world stage in 1956, after hostilities had erupted between Egypt and its adversaries France, Israel, and Great Britain over control of the Suez Canal. Both sides soon wanted to stop their conflict but could not bring themselves to take the steps needed to cease fighting and withdraw from the battlefield. Pearson proposed deploying unarmed or lightly armed forces from neutral countries, under United Nations command, to supervise the withdrawal of the warring forces and enforce the terms of the ceasefire. The first United Nations Emergency Force was born. The eventual and unequivocal success of the unassuming Canadian's idea won him the 1957 Nobel Peace Prize. In its citation, the Nobel selection committee deemed Pearson "the man who saved the world." And the world agreed.

SHROUDED TUYERE
The other bathtub discovery.

Admit it: you've broken wind while lounging in the bathtub. Robert Lee was one who 'fessed up. Cheerfully. It's how the Montreal metallurgist came up with the idea of using hot blasts of air to mix molten steel in what became known as a shrouded tuyere. Prior to Robert's 1958 all-natural bubble bath, steel was made by physically stirring the molten mixture as it sat in a forge or furnace. The process was inefficient, as well as being downright dangerous to the people who did the stirring. Installation of Robert's shrouded tuyere – usually made of copper – at the bottom of the furnace allowed hot air – sometimes enriched with pure oxygen – to enter and rise through the liquid metal, mixing it as the forced air rose. It's a life-saving device that comes from some careful observation, a dash of common sense, and, well, you know. Move over, Archimedes, your bathtub is big enough for two!

CRASH POSITION INDICATOR
The silent beacon.

What's more difficult than looking for the proverbial needle in a haystack? Searching for a downed aircraft in the rugged expanse of Canada's North. Even if a doomed flight emits a distress signal, finding its remains and survivors in a vast and inhospitable region is a chancy business. Many systems were tried and all failed, as multiple moving parts made these experimental contraptions uniformly untrustworthy. How could they survive the punishing force of impact? Harry Stevenson knew that a reliable device would have to consist of no moving parts whatsoever. With the support of Canada's National Research Council, the engineer created a tiny package with three fused parts: a transmitter, antenna, and delivery system. Mounted on the outside of an aircraft and launched by a spring-loaded mechanism upon impact, Harry's crash position indicator would be jettisoned a safe distance from the downed plane. The device was also waterproof, fire-resistant, and able to withstand even the biggest of jolts. As important, the antenna could transmit a signal regardless of its eventual orientation. The crash position indicator was an immediate success in 1959 and is now found on all commercial aircraft as part of the celebrated "black box" flight recorder. It is the silent beacon that leads us to that needle in the haystack.

GOALIE MASK

Innovation and common sense.

Joseph Jacques Omer Plante will be remembered as either the greatest innovator in hockey or just the most sensible goaltender ever to play the game. As a child, he played Canada's favourite sport in winter boots with a tennis ball and a goaltender's stick carved from a tree root. His first goalie pads were potato sacks wrapped around wooden slabs. Later, when his phenomenal talent brought him into the NHL as a Montreal Canadien, he showed up at every practice wearing toques he had knitted himself. By 1959 he was one of the most popular hockey celebrities in the history of the sport, but his face had taken a beating. In his spare time he devised a homemade solution, but his coach, Hector "Toe" Blake, forbade him to use it. At

least until Jacques threatened to strike. Minutes after being stitched up to close a gash caused by a flying puck at a game against the New York Rangers in 1959, a bloody Jacques said to his coach: "Let me play wearing my mask or I don't play at all." Although Blake argued that the mask would be taken by fans as a sign of cowardice, he had no choice but to accept. Moments later, Plante headed onto the ice wearing the device of his own creation – a moulded fibreglass mask that fit the contours of his face, with crude holes hacked out for the eyes. Sliding back into his position between the pipes, face mask on, Jacques changed his sport forever. From that day forward, no puck-stopper could imagine playing the game without a mask of some kind. Jacques Plante and his beloved Canadiens went on to win the Stanley Cup six times, and the innovator in the mask became a hero and role model to all.

REVERSE OSMOSIS
The salty tale.

Call it the irony of innovation: Canada is blessed with vast reserves of fresh water, yet a Canadian is responsible for creating a process to transform salt water into fresh water. His name is Srinivasa Sourirajan. In 1959, the Ottawa scientist — together with Sidney Loeb, a colleague at UCLA — designed a semi-permeable cellulose acetate membrane to desalinate water. The membrane is used to remove saline ions as pressurized water passes through it. The entire process is known as reverse osmosis. Within five years, asymmetric membranes were created to make the process commercially viable. Today, reverse osmosis membranes are also being used to treat hard water, clean polluted water, and recycle waste-water, and are being applied in all kinds of industries — as varied as medicine and maple syrup. Maybe instead of irony, it's just fair play.

THERMOFLOAT COAT
The mariner's guardian angel.

Whenever anyone died in the water near John Hayward's hometown, coroners most often ascribed the person's death to drowning. Hayward saw things differently. The University of Victoria thermal biologist surmised — and later proved — that hypothermia and not drowning was often the cause of a watery death, especially when the tragedy struck in the cool waters of Canada's Pacific coast. John Hayward and his team then acted on this knowledge. Together with research partners John Eckerson and Martin Collis, Hayward designed the first thermofloat coat. Taking a conventional floater jacket, the University of Victoria team added extra padding in areas where they had learned the body is most prone to losing heat — the neck and sides of the body. They also built in a removable flap that could be worn as shorts to reduce heat escaping from the groin and prevent cold water flushing up over the torso. Hayward and his partners sold manufacturing rights for their patented product to a small local apparel maker, and the mariner's guardian angel took flight.

BOMB SNIFFER
The smell of danger.

Did you know undetonated explosives give off faint traces of chemical vapour? Lorne Elias used this knowledge to create the first bomb sniffer in 1984. A researcher at Canada's National Research Council, he called his new nose the Explosives Vapour Detector, or EVD-1. The portable suitcase-sized device enabled airport security personnel around the world to check luggage quickly and reliably for a wide variety of explosives beyond dynamite. And it worked. During the 1984 papal visit to Canada, the device warned security teams that something was amiss in Pope John Paul II's luggage. Turns out that one of the pope's bodyguards had packed a revolver. Following the Air India bombing of 1985 in which 329 passengers en route to Bombay were killed, Lorne Elias's innovation became standard equipment at every Canadian airport. The success of the EVD spurred Canadian scientists to create an even faster and more reliable bomb-sniffing technique using ion mobility spectrometry to detect explosives by analyzing the electrically charged particles within them and identifying their unique toxic odours. Better security needs a whole new sense of smell.

PEALESS WHISTLE
The reliable alarm.

How many times did Ron Foxcroft blow his whistle to call a player for charging or travelling or committing a personal foul without being heard? Enough to spur the basketball referee to create a new kind of whistle. Ron spent three years working on his pealess version of the classic whistle. By 1987, he had perfected the device. It features resonant chambers that produce a trilling sound of many frequencies. Conventional whistles use pea corks to generate sound. But as Ron often found, cold temperatures or water or dirt inside these whistles caused them to fail. The pealess whistle is now found everywhere – around the necks of basketball referees, in the hands of hockey officials, within easy reach of rock climbers and snowboarders; it's even attached to life jackets. Call it innovation due to frustration.

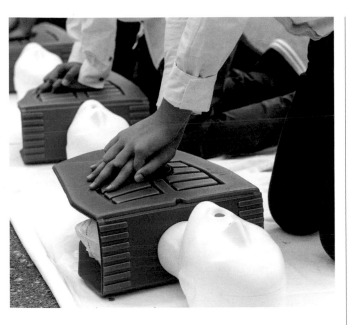

ACTAR 911
The life-size lifesaver.

Dianne Croteau has likely saved more lives than any other person, albeit indirectly. One of Canada's foremost industrial designers and inventors, she developed the ACTAR 911. The life-size yet light and easily transportable torso has been the industry standard for CPR training since she created it in 1989. The ACTAR 911 replaced the heavy, impractical mannequins used for years to teach cardiopulmonary resuscitation. Dianne's practical device has allowed many people to take CPR training who otherwise wouldn't have. And how many lives have these people saved?

ECOTRACTION
The road salt that doesn't kill.

Road salt kills. Many cancers in dogs and cats have been linked to its use. When a dog or cat licks its paws after a walk on a slushy street or sidewalk, the animal often ingests lead, mercury, and other cancer-causing toxins found in road salt. This staple of wintertime life is also harmful to humans. It sinks into the water table and, once there, can never be removed. Moreover, road salt inflicts untold damage on cars, roads, and bridges, making them less safe to use and more costly to maintain. Mark Watson was inspired by the death of his beloved cocker spaniel Grover to find an alternative to road salt. In 2005, the entrepreneur and his colleague Marc Appleby discovered a non-toxic, all-natural mineral that's a member of a special group of volcanic minerals called tectosilicates. Mark and Marc called it EcoTraction. The product embeds in ice and snow to create a non-slip surface. Yet unlike salt, it's harmless to plants, animals, and people, and it doesn't corrode cars or roads or bridges or anything.

HOLISTIC AIRCRAFT INSPECTION
The crystal ball of aeronautics.

Innovation often needs do nothing more than take a technique created for one purpose and suit it to another. That's exactly what Canada's National Research Council and Windsor, Ontario's Diffracto Limited accomplished together beginning in 1990. The company produces a tool that uses light to inspect automobiles and reveal any damage and distortion invisible to the naked eye. The two organizations adapted the method to sweep large areas of aircraft in search of imperfections. The technique not only speeds the inspection process, but also improves it by enabling aeronautical technicians to focus their time and energy looking beneath the surface at specific, targeted places for evidence of damage caused by age, corrosion, and impact. This crystal ball of aeronautical engineering has dramatically changed the way all aircraft around the world are inspected, maintained, and repaired. In the process, this change has saved money, reduced risks, and increased safety. It has proven so valuable that, even though it was developed initially for use on aircraft, it's now influencing how bridges and other types of public infrastructure are designed and built. Purpose, repurpose, and repurpose again.

MIOVISION
The smarter traffic counter.

They are perhaps the surest sign that summer in the city has arrived: those young women and men who sit in lawn chairs near intersections and count vehicles. One of the stalwarts performing these summertime summations was Waterloo, Ontario's Kurtis McBride. He found the job tedious for the counters and a wholly inefficient way for municipalities to gather traffic data. Rather than just gripe, he developed Miovision. The traffic-management technology captures and then processes millions of hours of traffic video into highly accurate data. Since 2005, municipal governments around the world have relied on his advance to improve transportation networks and make sure city infrastructure meets today's needs. It's a case of one smart counter building an even smarter one.

HERSWAB
The self-test for cervical cancer.

Cervical cancer claims more than 275,000 lives each year, mostly among women who fail to undergo screening for the virus that causes the disease. The problem isn't technical; it's behavioural. Many women don't get tested for high-risk human papillomavirus because they don't have the time, lack access to doctors, find the test embarrassing or uncomfortable, or are inhibited culturally. When added up, the numbers are staggering: fully one-third of women in the world's developed countries either have never screened for the virus that causes cervical cancer or don't get screened regularly. That promises to change. In 2010, Jessica Ching, an industrial designer in Toronto, created HerSwab. The world's first self-test for cervical cancer, it's designed to enable women themselves to collect samples near their cervixes, where traces of the human papillomavirus are most likely to be found. The self-test for cervical cancer is here. Now let's see what it can do.

HELICOPTER CUSHION
The safer ride.

One grave threat to helicopter safety comes from an unlikely source – pilot discomfort. The continual shaking and jolting that helicopter pilots must endure, often for hours at a time, has serious consequences. It not only causes soreness and fatigue – both safety threats – but also leads to a variety of long-term physical ailments such as chronic pain in the back and neck. In 2011, researchers from both Canada's Department of National Defence and National Research Council used their vast knowledge of the causes and effects of vibration to develop a new helicopter cushion. Its design integrates traditional foam with an energy-absorbing material. The hexagonal cell pattern of the material blends with a system of air vents to dissipate vibrational energy. Most importantly, the new cushion enhances the comfort of pilots without compromising the airworthiness or safety of the craft. In fact, a more comfortable ride is a safer one. The innovation is now being integrated into the new armoured seat for the Canadian Armed Forces CH-146 Griffon to help improve the comfort and health of its aircrew.

NO-REPAIR BRIDGES
The concrete solution.

The deck of the Canal Bridge won't need to be resurfaced until well into the twenty-second century. It's not magic. Well, maybe just a bit of magic. The cross-border bridge in Cornwall, Ontario – finished in 2016 – is covered with a new class of concrete. Coming up with just the right mix of ingredients involved a lengthy experimenting and testing process of many sophisticated concrete technologies and precise proportions of cement, water, chemicals, special sands, and coarse aggregates. Inspired by Canada's Critical Concrete Infrastructure project, the team behind the mixing is just as complex – an intricate combination of suppliers, government agencies, and research organizations including FBCL, Transports Québec, National Capital Commission, City of Ottawa, W.R. Grace, Lafarge North America, and Northeast Solite. The result is a concrete that enables bridges to last some five times longer than ones made from the conventional hard stuff. Yes, the new mix costs about 20 per cent more, but the cost savings for bridge owners and taxpayers generated by the strength, durability, and safety of no-repair bridges will more than offset the higher price tag. The bridge to the future is here, and it's being built in Canada.

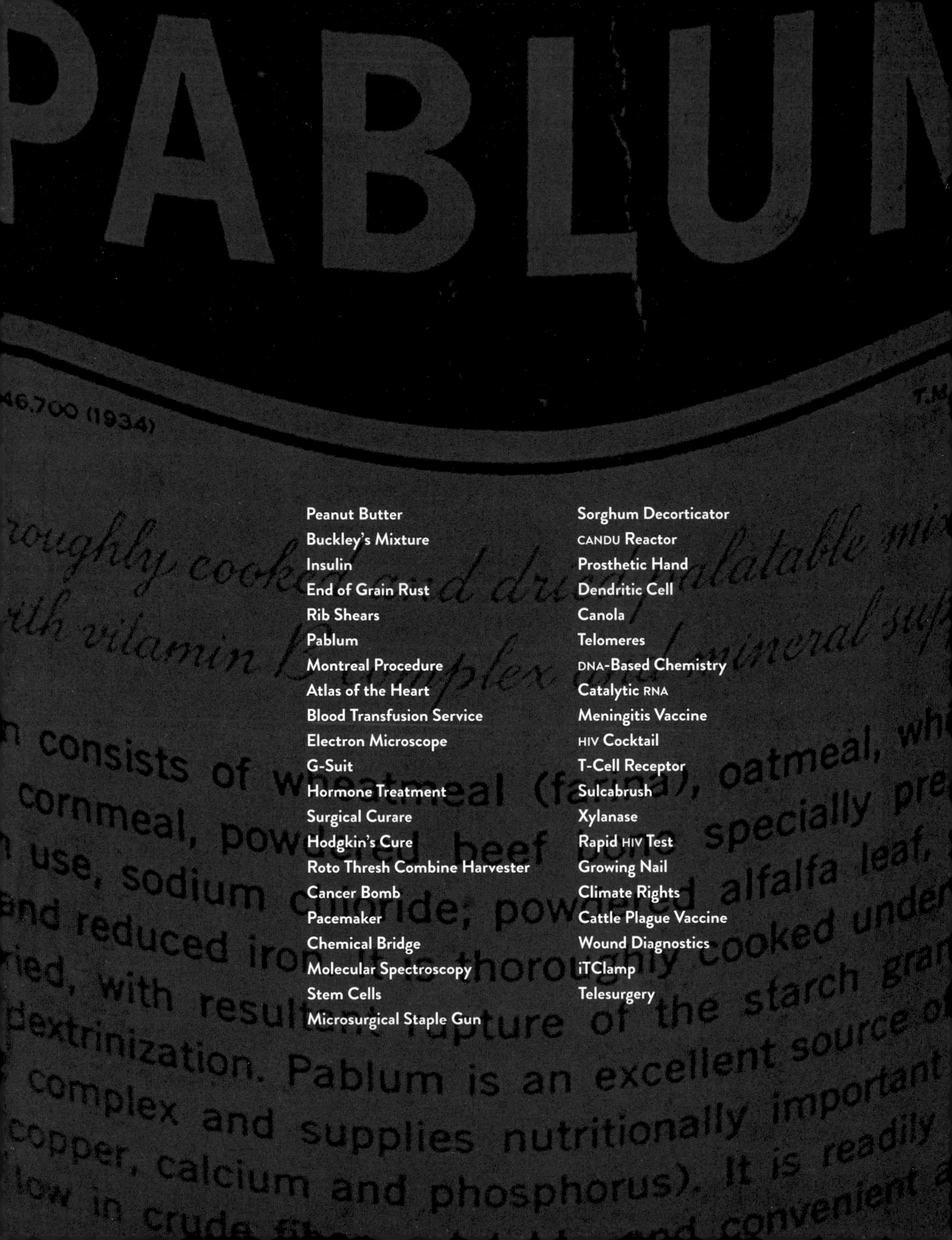

PABLUM

46,700 (1934)

T.M

roughly cooked and dried palatable mix
with vitamin B complex and mineral sup

consists of wheatmeal (farina), oatmeal, whe
cornmeal, powdered beef specially pre
use, sodium chloride, powdered alfalfa leaf,
and reduced iron. It is thoroughly cooked under
ied, with resultant rupture of the starch gran
dextrinization. Pablum is an excellent source of
complex and supplies nutritionally important
copper, calcium and phosphorus). It is readily
low in crude fiber... and convenient

Peanut Butter
Buckley's Mixture
Insulin
End of Grain Rust
Rib Shears
Pablum
Montreal Procedure
Atlas of the Heart
Blood Transfusion Service
Electron Microscope
G-Suit
Hormone Treatment
Surgical Curare
Hodgkin's Cure
Roto Thresh Combine Harvester
Cancer Bomb
Pacemaker
Chemical Bridge
Molecular Spectroscopy
Stem Cells
Microsurgical Staple Gun

Sorghum Decorticator
CANDU Reactor
Prosthetic Hand
Dendritic Cell
Canola
Telomeres
DNA-Based Chemistry
Catalytic RNA
Meningitis Vaccine
HIV Cocktail
T-Cell Receptor
Sulcabrush
Xylanase
Rapid HIV Test
Growing Nail
Climate Rights
Cattle Plague Vaccine
Wound Diagnostics
iTClamp
Telesurgery

Smarter Smaller Kinder Safer Healthier Wealthier Happier

If we have only our health, we're all in better shape thanks to Canada. With long and deliberate collaboration among the creative teams found in hospitals, universities, research agencies, private corporations, and national institutions bent on innovation, Canadians have found cures and treatments for diseases that include diabetes, meningitis, Hodgkin's disease, cattle plague, and AIDS. They have built and devised instruments such as electron microscopes, rib shears, cobalt radiation machines, surgical staple guns, and wound clamps that have enabled new life-saving treatments. And they've invented healthier foods such as Pablum and peanut butter that have boosted nutrition, prevented early death, and extended life.

PEANUT BUTTER
The protein substitute.

Step aside, George Washington Carver. Contrary to almost universal belief, the celebrated American botanist didn't create peanut butter. The stick-to-the-roof-of-your-mouth glory goes to Marcellus Gilmore Edson. In 1884, the Quebec chemist was awarded the first patent for peanut butter—or peanut-candy, as it was called then. Marcellus discovered it when he found that heating the surfaces to grind peanuts to 100 degrees Fahrenheit caused crushed peanuts to emerge as a thick, chunky fluid. When the liquidy grounds cooled, they set as a paste similar to butter. Now enter John Kellogg of the cereal empire, who marketed the creamy spread as a protein substitute for people who couldn't eat solid food. Sorry, Mr. Carver.

BUCKLEY'S MIXTURE
The mouthful of brisk.

A cough in the spring of 1919 was no trivial matter. The flu was a deadly threat, not a temporary annoyance. When customers of William Buckley's Toronto pharmacy came to him seeking a stronger cough suppressant than others then available, he could offer no help. That bothered him. Buckley quickly got to work and came up with a blend of menthol, pine-needle oil, ammonium carbonate, and Irish moss extract. It suppressed coughing all right but tasted, um, unique. "Brisk" is how he characterized it. To desperate patrons eager for relief, it was a godsend, regardless of its taste. Buckley's mixture sold just as briskly as it tasted, prompting the pharmacist to open his own manufacturing plant. It may not have saved lives, and brisk might just be a polite word for horrible, but it worked. Still does.

INSULIN
The end of terror.

Diabetes. There was a time when the word struck terror in the hearts of parents. A child diagnosed with the disease could expect to live a life of perpetual illness and suffering that would likely end in death before the child emerged from adolescence. That the word no longer strikes terror is largely because of three Canadians: medical scientist Dr. Frederick Banting, his assistant Charles Best, and their University of Toronto patron and adviser J.J.R. Macleod. In their historic summer of 1921, Banting and Best isolated what we now know as insulin while performing experiments in a laboratory loaned to them by the vacationing Macleod. Over the next several months, the small team, which now included biochemist James Collip, refined their discovery until it proved a reliable remedy for the dread disease. The drug firm Eli Lilly and Company came on board in late 1922, using its corporate resources to produce large quantities of highly refined insulin, thereby making the treatment available widely and ending the terror.

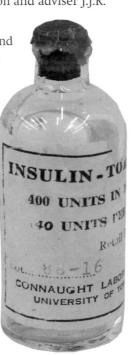

END OF GRAIN RUST
The fight with blight.

Each of us could only hope to enjoy the definitive professional success of Margaret Newton. In 1925, Canada's minister of agriculture appointed her to manage the newly opened Dominion Rust Research Laboratory at the University of Manitoba and gave her the task of defeating grain rust. At the time, this pathogenic fungus was a plague of the nation's harvest, destroying some thirty million bushels of wheat each year. When she retired some twenty years later, that figure was zero. All credit to Dr. Newton. During her career, she wrote more than forty research papers on the subject of rust fungi, a wealth of knowledge that government officials and farmers used to eradicate the blight from Canadian crops. The fight was won.

RIB SHEARS
The fastest way to the heart.

Dr. Norman Bethune was hard to peg: selfless hero or egoist, lover of humanity or cruel friend and husband, brilliant medical mind or carelessly impatient surgeon. What is unmistakable about the man's legacy is the enduring contribution of the surgical tool that bears his name. The Bethune rib shears — designed and developed by the thoracic surgeon during his time at Royal Victoria Hospital in Montreal in the early 1930s — feature a long s-shaped handle and small head. Both attributes are ideal for precise clipping through the delicate bones, muscles, and connective tissue or fascia that shield the organs within the thoracic cavity. Strong yet delicate, they are also ideal for cutting through the ocular bone during eye surgeries. Innovative, simple, and practical, Dr. Bethune's shears are still used around the world more than eighty-five years after he created them.

PABLUM
The other infant nutrition.

All kinds of ready-made foods for infants, regardless of their actual brand names, are called Pablum. This fact attests to the extraordinary success of three Toronto paediatricians. Spurred by rising numbers of malnourished children, Drs. Alan Brown, Fred Tisdall, and Theo Drake of the Toronto Hospital for Sick Children developed a formula for infants and youngsters. Pablum had two advantages over other food products for children: it was easy to make and, packed with minerals and vitamins, it nourished anyone who ate it. The impact of Pablum was immediate. Soon after introducing the formula in 1931, hospital officials saw mortality rates of infants and young children plummet. Pablum was not only a boon to kids; the hospital itself grew stronger. When negotiating the rights to produce and sell Pablum, Dr. Tisdall made sure the hospital retained patent rights to the formula and, by doing so, secured millions of dollars in royalties to support paediatric research. As it turned out, Pablum markedly improved both infant and institutional health.

MONTREAL PROCEDURE
The wakeful brain surgery.

Don't read the following words if you get queasy easy. They are about what is known as the Montreal procedure. Developed in 1931 by Dr. Wilder Penfield of the Montreal Neurological Institute of McGill University, the procedure is a neurological method to eliminate severe epileptic seizures. It goes like this. The neurosurgeon administers a local anaesthetic to the patient, keeping him or her conscious during the procedure. The surgeon then removes a piece of the patient's skull to expose the brain tissue. While the surgeon probes the brain, the patient describes what he or she feels so the surgeon can pinpoint the exact location of seizure activity in the brain. The surgeon then destroys the nerve cells in this spot of the brain to bring the epileptic seizures to an end. Dr. Penfield found that more than half of the patients he treated with the Montreal procedure were cured of seizures. The method also enabled him to create maps of the sensory and motor cortices of the brain and show vividly how the brain connects to the limbs and organs of the body. These maps, virtually unchanged, are still in use today. Okay, the coast is clear. You queasy types can start reading again.

ATLAS OF THE HEART
The cardiac catalogue.

To fix something, you first must have a decent understanding of what's wrong with it. So to have any hope of repairing a damaged heart, you first must know all the ailments and abnormalities that can afflict this vital organ. *The Atlas of Congenital Cardiac Disease* gave heart surgeons that critical knowledge for the first time. Published in 1936, the cardiac catalogue was the work of a remarkable woman named Maude Abbott. A trained medical pathologist, she used her position as assistant curator of McGill University Medical Museum to collect and study the hearts of people who had died of cardiac problems. A tireless professional, she also scoured historical records, scrupulously compiling cardiac anomalies identified during autopsies. When finished, her work became the foundational document of modern heart surgery, stimulating multiple advances in the physiology of the organ and the diagnosis of heart ailments. What makes Dr. Abbott's accomplishment all the more remarkable is the sexism she encountered at every stage of her career, beginning with being denied entry to medical school at McGill because she was a woman, a practice that ended finally in 1918. Call it a long-overdue change of heart.

BLOOD TRANSFUSION SERVICE
Old method meets new need.

Innovation often emerges when an old method meets a new need. Such a meeting occurred in Spain in 1936. The man at its centre was Norman Bethune. The Canadian physician had journeyed across the Atlantic to serve Loyalist forces in that country's civil war. Unable to find a post as a surgeon, Bethune found another way to contribute. He set up a mobile transfusion service that transported blood donated by civilians to treat wounded soldiers near or on their fields of battle. Dr. Bethune was likely inspired to create a blood transfusion service by his experience administering transfusions during the surgeries he performed at Montreal's Sacred Heart Hospital.

His courageous innovation was taking the service right into the theatre of war where it was so widely and urgently needed.

Bethune was not an easy sort to get along with and, weary of his abrasive personality, some of his Spanish comrades suggested he move on. His blood transfusion service lived on – in Spain and soon wherever else battles raged. Less than two years later he was on the battlefields of China, acting as medical adviser and surgeon to combatants on both sides of the communist revolution. Much admired for his innovative spirit and high ethics by Mao Tse-tung, Norman Bethune was the only foreigner mentioned in *Quotations from Chairman Mao Tse-tung*, the Little Red Book that became one of the most printed books in history.

ELECTRON MICROSCOPE
The first electric eye.

There's more to it than meets the eye. That phrase wasn't just an idiom for James Hillier and Albert Prebus. In 1938, the two University of Toronto Ph.D. students took advantage of rapidly emerging knowledge of the inner workings of the atom to create the electron microscope. While conventional optical microscopes relied on light waves to magnify objects, Prebus and Hillier's device worked by focusing a beam of electrons. The wavelength of electrons is much smaller than the wavelength of light, making it possible for the electron microscope to achieve much greater magnification and depth of focus. The students' new device magnified objects to seven thousand times their size, compared to the two thousand achieved by traditional optical microscopes. The two weren't done; within a few years, they had created a scanning version of their device that works by scanning an object with a beam of electrons to generate an image that is then visible on a screen. Scientists capitalized quickly on the advance, looking more deeply and clearly to gain new insights

into biology, medicine, and materials. Electron microscopes remain fixtures in research labs around the world, proving daily that there is much more to the world than meets the eye.

G-SUIT
The guarantee of grace under pressure.

Advances in technology can lead to unintended consequences. As Allied aircraft in World War Two became faster and capable of reaching higher altitudes, pilots began suffering temporary hypoxia, a medical term meaning they passed out. Pilots blacked out because the increase in gravitational pressure on their bodies was drawing oxygen-rich blood away from their brains and into their legs. Enter Wilbur Franks. In 1940 the pioneer in aviation medicine at the University of Toronto worked with his team to design a flight suit equipped with water-filled bladders. When the pilot experienced an increase in gravitational force, the bladders inflated. This pressure on legs and midsection enabled a pilot's blood to circulate normally. The Franks Flying Suit, and then his Mark I and II models, was put into use immediately by British, American, and Canadian air forces. Later designs replaced water pressure with that of air. They still enable pilots and even astronauts to withstand the unintended consequences of high-speed travel.

HORMONE TREATMENT
The new way to battle cancer.

Charles Huggins had a clear approach to his work. "Discovery is for the single mind, perhaps in company with a few students. Don't write books. Don't teach hundreds of students. Make damn good discoveries." The Halifax-born physician's research made good on his pledge. Through his work, he showed that cancer cells are not autonomous and self-perpetuating. They are dependent on hormones and other chemical signals to grow and survive. In 1941, he discovered that hormones could be used to control the spread of some cancers, such as prostate. Ten years later, he found that breast cancer, like prostate cancer, was dependent on specific hormones and that advanced breast cancer could be influenced through hormonal manipulation. In 1966, Dr. Huggins was awarded the Nobel Prize in medicine for his work in the hormonal treatment of cancer – a good discovery.

SURGICAL CURARE
The upside of poison.

Medical historians consider the field of anaesthesia to fall into two eras – before Griffith and after Griffith. The Griffith in question is Dr. Harold Griffith, chief of anaesthesia at the Montreal Homeopathic Hospital. In 1942, he became the first person to use curare – which until then had been considered strictly a poison – as a surgical anaesthetic. His move had breathtaking consequences on surgery. It proved a careful dose of the muscle relaxant could dramatically and safely reduce the reliance on conventional methods of anaesthesia. It increased the scope of surgery. It improved operating conditions. It decreased the likelihood of disease and death resulting from surgery. And it paved the way for development of dozens of anaesthetics, many of which are still relied on in hospitals around the world – the after-Griffith world.

HODGKIN'S CURE
The two fights.

Vera Peters fought two battles. Her first involved overcoming the sexism of her time to graduate from medical school in the 1930s and emerge as an expert in oncology. Her second was against Hodgkin's disease. In 1950, Dr. Peters discovered that the illness – a cancer of the lymphatic system considered terminal – could be cured in its early stage through a treatment of high doses of radiation. Her colleagues were sceptical. Yet another battle for her to wage and eventually win. Dr. Peters's treatment worked consistently and was soon used to cure breast cancer as well. She was one of many pioneering Canadian women in medicine and science who, while overcoming the needless challenges imposed on them by their society, solved some of the most daunting scientific problems of their time. They fought two fights.

ROTO THRESH COMBINE HARVESTER
The separator of grain and chaff.

Combine harvesters changed little over 150 years until two Manitoba farmers gave this vital machine a new spin. William and Frederick Streich's roto thresh combine harvester was the first harvesting machine to use the centrifugal force of a spinning drum to reap and separate grain from chaff and straw. Conventional combines relied on gravity to achieve the separation. The pair's new spin enabled farmers to harvest more grain and lose fewer kernels than in the threshing cylinders of traditional machines. They developed their prototype in 1950, with the first production model finally rolling off the assembly line almost twenty-five years later in 1974. Only fifty were built, however. Emerging knowledge about the value – or lack thereof – of pulverized straw to restore soil fertility meant farmers returned to the more conventional combines, which left behind a higher quality of straw that could be baled, removed from fields, and used for livestock fodder. Fresh insight about soil put a new spin on the new spin.

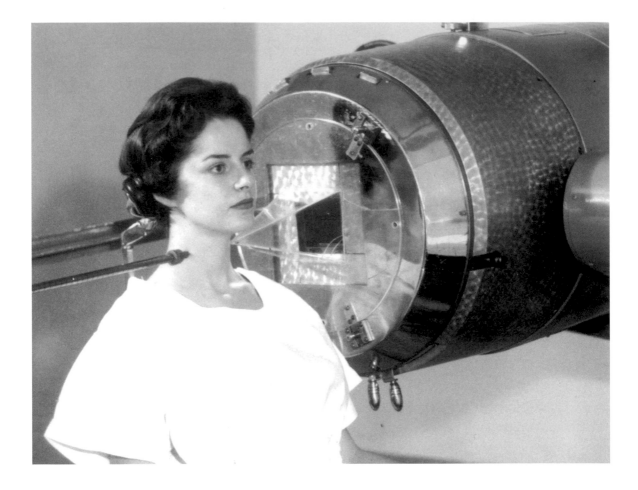

CANCER BOMB
The advanced treatment.

Use something deadly to eliminate something even deadlier. That was the inspired thinking behind the world's most successful cancer treatment. Developed in 1951 by Drs. Sylvia Fedoruk and Harold Johns, it directed deeply penetrating gamma radiation from radioactive isotope cobalt-60 at malignant tumours. The cobalt radiation unit set up by the two University of Saskatchewan researchers in the 1950s treated some 6,700 patients, saving or prolonging many of their lives and giving hope to countless others since. In its most famous iteration, beginning in 1957 the radiation was delivered by the Theratron Jr., which fired gamma rays directly at and deeply into the afflicted area of the body. The pioneering researchers' cancer bomb is still relied on throughout the world as a chemotherapy treatment for cancer. That's one deadly beneficial bomb.

PACEMAKER
The quicker ticker.

Unravelling the mysteries of the human heart is not just for poets and songwriters, but also physicians and engineers – especially Canadian ones. Starting in the 1940s, physician Wilfred Bigelow experimented in Toronto with extreme cold as a way to slow the rate of patients' hearts during open-heart surgery. It was a promising method, but he couldn't figure out how to safely restart a heart that had stopped beating. Engineer John Hopps at Canada's National Research Council was tasked with finding the solution. By 1951, he had developed an electrical pacemaker circuit that supplied a gentle stimulation to the heart. The charge came from a metal cabinet the size of a cereal box that used vacuum tubes to generate electric pulses. An insulated wire from the cabinet was inserted through a vein in the neck and then snaked down to the heart. A dial controlled the power of the pulse, mimicking the organ's natural impulse without injuring the heart muscle. Hopps's achievement sparked a wave of research in biomedical engineering and placed Canada at the forefront of the field. By 1965, the country's National Research Council had developed the world's first biological cardiac pacemaker. Advanced versions eventually found their way into the bodies of millions of people around the world. One of the millions was John Hopps. Engineer, heal thyself.

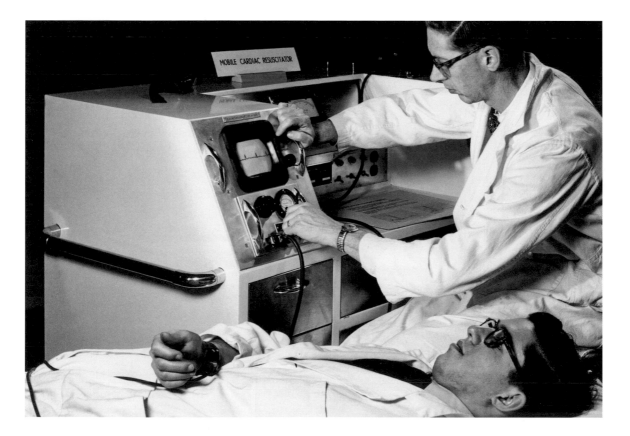

CHEMICAL BRIDGE
The other electron transfer.

Reflecting on his early career, Henry Taube revealed he had trouble finding graduate students to take part in his work. They didn't find his study of electron transfer reactions to be exciting enough. Boy, were they wrong. The Canadian-born chemist is universally recognized as the founder of the modern study of inorganic mechanisms. His most important discovery, dating from 1952, is the way molecules build chemical bridges instead of simply exchanging electrons as conventional wisdom had suggested. This intermediate step in the process explains why reactions between similar metals and ions occur at different rates. For this work, Dr. Taube was awarded the 1983 Nobel Prize for chemistry. The award didn't change the humble, unassuming chemist. But he did find it changed his students; they took a new-found interest in his exciting work.

MOLECULAR SPECTROSCOPY
The deep view.

Peer deeply into a molecular phenomenon that lasts for a few millionths of a second. That was the daunting challenge that Ottawa physicist Gerhard Herzberg posed to himself in 1959. Dr. Herzberg concentrated his sights on methylene, an unstable molecule known as a free radical. Free radicals are created ever so briefly in a chemical reaction when molecules come together and then rearrange themselves to create something new. Working with a team of colleagues, the good doctor applied his formidable talents and identified the microwave, infrared, and ultraviolet regions of the electromagnetic spectrum of this free radical. His was an extraordinary achievement—not only for what it told us about the structure of this fleeting molecule, but also what this knowledge has since enabled others to achieve in other fields, notably medicine and advanced materials. By peering deeply, Dr. Herzberg made it possible for others to see further, and his Nobel Prize proved that the world admired the view.

STEM CELLS
The matter with marrow.

A strange thing happened when, during a 1961 radiation experiment, James Till and Ernest McCulloch increased the amount of bone marrow they injected into mice. Two things actually. They found the injections increased the survival rate of the irradiated mice. They also noticed odd lumps on the spleens of the mice. Further investigation revealed the lumps to be cloned cells caused by the injections. Even more study by physician McCulloch and physicist Till determined the cloned cells, which they dubbed stem cells, had the ability to renew themselves and could also differentiate into more specialized cells. These observations have since opened up a new field of medical research and promise to hold open a range of treatments for diseases, ailments, and injuries once thought untreatable. Strangely wonderful things can happen when you keep your eye on the straight and marrow.

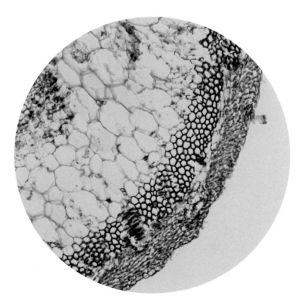

FIND INVESTORS
FOR YOUR IDEA: HERE'S HOW.

☐ Know that the best source of funding are the customers who pay for your product. Only when you have to grow fast, acquire more customers, build more tech, hire more people, and cement partnerships should you need investment capital.

☐ Find a mentor, and join a business accelerator where you can learn how to attract both customers and investors.

☐ Refine your investor pitch by asking entrepreneurs in residence to poke holes in your value proposition.

☐ Identify the type of investor you want, and consider what they might provide other than money, including mentorship, contacts, customers, technology, credibility.

☐ Talk to companies that got investment from your target investors. Get feedback on the style and preferences of each. Find out what works best in each case.

☐ Engage a corporate legal firm to structure your company properly for investment. Ask if they have a start-up package.

☐ Be prepared to give a reasonable percentage of your company to your investors. It's not bad to have only a small percentage of what might become a huge success.

☐ Know exactly how you will use the investment funds to get to a defined milestone.

☐ Be upfront when you pitch. Answer every question directly. Never exaggerate. Make your projections conservative. Acknowledge that you have competitors. Admit when you don't know something, and pledge to get back to them later.

☐ Get used to being rejected. Investors' perspectives are different from yours. Learn, adapt, persist.

MICROSURGICAL STAPLE GUN
The tiny closer-upper.

George Klein was Canada's most prolific innovator of the twentieth century. Yet in a life rife with engineering achievement, perhaps his most consequential creation is the microsurgical staple gun. Instead of fastening plywood to struts, this little beauty seals minute blood vessels of the human body. To create it, Klein joined a team of Canadian doctors that included famed surgeon Isaac Vogelfanger. Together in 1962 they developed a device formed of two parts. The first, like forceps, surrounds the ends of the severed vessels and draws them together. The second encircles the connection and, with numerous small metal staples, seals it with a quick snap. Faster and better than traditional suturing, the team's microsurgical stapling device saved many lives. Its greatest contribution, however, may have been in early kidney transplant and heart procedures that would have been impossible without the device's rapid precision.

SORGHUM DECORTICATOR
The developing world's huller.

Few devices have had a greater influence on the lives of people in the developing world than a humble machine from Canada. The sorghum decorticator processes sorghum, millet, and other staple crops grown by farmers in India and throughout Africa. What's so special about this work? These grains cannot be eaten at all until the outer layers are completely removed, and the machine can hull them in a fraction of the time it once took women and children to do it. Developed in 1970 by Canadian engineers, the sorghum decorticator is small, portable, and fast, making it possible for farmers and those in their communities to feed themselves rather than rely on expensive imported grains such as rice and wheat. A humble machine but no humble achievement.

CANDU REACTOR
The source of clean power.

While other countries girded for war, Canada prepared for peace. Nuclear scientists around the world spent most of the years following World War Two exploring ways to build more powerful weapons. Yet researchers at Atomic Energy of Canada committed themselves to investigate ways to produce and deliver nuclear energy for peaceful use. George Laurence was the chief figure in the search. His work led to development of the CANDU nuclear reactor in 1971. CANDU stands for Canada Deuterium-Uranium. Deuterium (in the heavy water) and uranium (the fissile material) are the two most important elements in the reactor. The pioneering structure generates electricity by using the energy created by the fission of uranium to produce steam. The steam turns a turbine that creates

electricity. What sets the CANDU apart is its safety and efficiency. The reactor core is made up of hundreds of fuel channels set in a grid. The channels pass horizontally through a tank that contains heavy water. The heavy water moderates and controls the buildup of energy from the fusion of the uranium. The water and channels are essential for fission to occur in the uranium. If the reactor is damaged in any way, either the makeup of the water or the structure of the grid changes, and the fission ceases immediately. Peaceful power: the gift of a peaceful nation.

PROSTHETIC HAND
The new reach.

Prostheses are almost as old as human civilization, dating to the days when Egyptians fashioned wooden toes and fingers for missing digits. Yet in the thousands of years since then, artificial hands and limbs never became much more sophisticated than hooks and pegs. World War Two forced many countries to rethink their attitude toward prostheses. Canada was at the forefront of the new thinking, establishing an artificial limb program in the final months of the conflict to equip injured veterans with prostheses that were more than limp appendages. Helmut Lucas was the program's most gifted designer and craftsman. In 1971, the medical scientist created a prosthetic hand that incorporated electronic and mechanical elements. These components made it possible for those equipped with the Lucas hand to perform tasks impossible with the crude pre-war prosthetics. Actions as seemingly simple as turning a doorknob, picking up a glass, and buttoning and unbuttoning a shirt were life-altering achievements for disabled veterans. Helmut's new hand put within reach a better life for these men and many other people to come.

DENDRITIC CELL
The engine of adaptive immunity.

"I know I have got to hold out for that. They don't give it to you if you have passed away. I have got to hold out for that." In the fall of 2011, Ralph Steinman was deathly ill with pancreatic cancer. The Canadian immunologist clung to life, awaiting word from Sweden on whether he would be awarded the Nobel Prize in medicine. In 1973, while working at Rockefeller University in New York, Dr. Steinman had discovered what he called dendritic cells. The cells are essential elements of the human immune system. Their main function is to process antigen material to make it present on the surface of a cell so that T cells can interact with the material. An antigen is a molecule capable of inducing an immune response on the part of a host organism. T cells are a subtype of white blood cells, which are central components of immunity in human beings. Since their discovery, dendritic cells have become known as the primary instigators of adaptive immunity and have been used to design vaccines to fight HIV and several forms of cancer. Alas, cancer claimed Dr. Steinman before he received word from Sweden. Fittingly, the Nobel committee granted him the award nonetheless. He had held out long enough.

CANOLA
The superior cooking oil.

Rapeseed was the cash crop for generations of Canadian farmers. The oil produced from the bright yellow–flowering member of the mustard family was used to lubricate the world's steam engines. But when diesel replaced steam, the demand for rapeseed plummeted, taking many Canadian farm incomes with it. Baldur Stefansson and Keith Downey found an alternative use for rapeseed oil. In 1974, the two agricultural scientists at the University of Manitoba carried out a series of cross-breeding experiments on rapeseed plants until they were able to create a version that had little erucic and eicosenoic acids – two acids that made rapeseed oil perfect for lubrication but awful for cooking. They called their new plant canola. The oil from this hybrid had a higher nutritional value and lower trans fats than almost every edible alternative, including butter and lard. Today, this distinctive Canadian oil is one of the most popular edible oils and the canola plant is one of the largest oilseed crops, not only for new generations of Canadian farmers but also for farmers throughout the world.

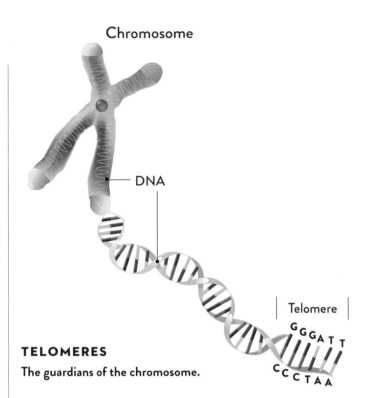

Chromosome

DNA

Telomere

GGGATT
CCCTAA

TELOMERES
The guardians of the chromosome.

Our chromosomes make us who we are. Each one is an organized structure that contains most of the DNA of a living organism. No one has contributed more to our understanding of these tiny packages of life than Jack Szostak. The University of British Columbia biochemist constructed the world's first artificial chromosome. This achievement helped scientists map the location of genes in mammals and develop techniques for manipulating genes. With colleagues Carol Greider and Elizabeth Blackburn, in 1975 and the two following years he clarified the events that led to chromosomal recombination, which is the reshuffling of genes that occurs during meiosis. And he discovered how chromosomes are protected by telomeres, which are specialized DNA sequences at the tips of chromosomes, and the enzyme telomerase. For his groundbreaking work in telomeres, Dr. Szostak was awarded the 2009 Nobel Prize for medicine. That's who *he* is.

DNA-BASED CHEMISTRY
The key to genetic disease.

Studying life requires more than knowledge of biology. Principles of physics and chemistry must also be considered to understand how the spark of life occurs and how organisms grow. For this insight, we can thank Michael Smith. In 1977, the Canadian researcher developed site-directed mutagenesis. It's a method to make precise and deliberate changes to the DNA sequence of a gene — the very chemical locus of life. Call it DNA-based chemistry. When researchers make these changes, they are able to investigate the structure and biological activity of DNA, RNA, and protein molecules — the fundamental building blocks of life. DNA chemists could even use Dr. Smith's method to manipulate genes and proteins to create new forms of life. For his groundbreaking work, he was awarded the Nobel Prize in chemistry, proving conclusively that life is more than mere biology.

CATALYTIC RNA
The secret of heredity.

Some young Canadians are entranced by sports or music or words. Growing up in Montreal, young Sidney Altman was enchanted by the periodic table. At the age of thirteen, he first saw in the table the elegance of scientific theory and its predictive power. Enchantment eventually led to revelation. Working with colleague Thomas Cech at Yale University in 1982, Dr. Altman discovered that RNA (ribonucleic acid) molecules are not restricted to being passive carriers of genetic data, which was conventional wisdom at the time. They also play active roles within cells by acting as catalysts in biological reactions, functions once thought to be carried out by enzymes only. The implications of this discovery cannot be overstated. One day, specially designed RNA molecules may be used to destroy harmful properties in certain organisms, including human beings. For their pioneering work, Drs. Altman and Cech were awarded the Nobel Prize in chemistry. Perhaps one day an even greater reward will come when that work inspires some other thirteen-year-old to appreciate the elegance of scientific theory and its predictive power. Perhaps it already has.

MENINGITIS VACCINE
The end of one childhood tragedy.

Meningitis is an infection of the fluid and lining of the brain and spinal cord. It strikes quickly and targets the most vulnerable; half its victims are under two years, and it kills up to 10 per cent of those afflicted within two days. Youngsters who survive are often left with permanent brain damage, hearing loss, and other grave health problems. Yet little has been heard of meningitis for some thirty years. Why is that? The answer is Dr. Harold Jennings. In 1982, the medical researcher and his Ottawa-based team concluded development of a meningitis vaccine. The vaccine causes complex sugars to combine and cover the Group C meningitis bacteria with a protein. On top of that, the vaccine is incredibly practical; each dose can be measured precisely, dissolved in solution for injection, and tolerated by infants. The impact of Dr. Jennings's discovery was swift and sure. Within two years, the United Kingdom, the first country to use the vaccine in a mass immunization program, virtually eradicated bacterial meningitis. The end of one man's work spelled the end of one childhood tragedy.

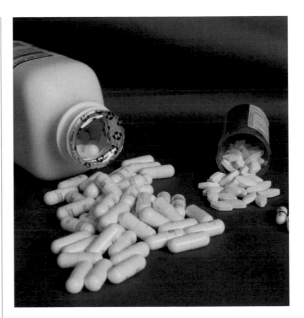

HIV COCKTAIL
The saviour of two million.

What greater gift can one person make than saving a life? How about two million? That's the legacy of Bernard Belleau. The Montreal research chemist created 3TC, a drug used to fight AIDS. Developed in 1983, it works by preventing the HIV retrovirus from altering the genetic structure of infected cells, thereby stopping the spread of the virus throughout the body because no new cells can be infected. It also serves as an alternative to AZT, one of the earliest drugs used to combat the dread disease. AZT, however, had fallen into disfavour: not only did it cause debilitating side effects, but patients also quickly grew immune to its anti-HIV properties. Don't toss out AZT just yet, though. Dr. Belleau and his colleagues, Gervais Dionne and Francesco Bellini, found that 3TC combined with low doses of AZT worked better than 3TC alone. The HIV cocktail was born, and two million lives were saved. Quite a legacy.

T-CELL RECEPTOR
The map to the Holy Grail.

When experts from different disciplines get together to tackle shared problems, innovation is sure to follow. Tak Wah Mak of Toronto's Princess Margaret Cancer Centre believes in this wisdom. It's how he's achieved his greatest break-throughs. In 1984, Dr. Mak assembled a large team of clinician-researchers to discover the T-cell receptor. The TCR, as it's known, is a molecule found on the surface of T cells, which play a central role in the immune system. He also says this approach enabled the team to develop a cancer-fighting drug. Known as CFI-400945, the drug targets an enzyme that plays a vital role in the division of cancer cells. Target is the key word. Unlike chemotherapy, Dr. Mak's sharp-shooter drug doesn't affect healthy cells. It makes cancer cells unstable and renders them more vulnerable to treatment and, perhaps, an ulti-mate cure, which is medicine's Holy Grail.

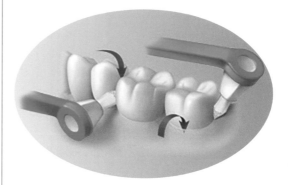

SULCABRUSH
The chewing stick reinvented.

Canadian dentist Dr. Max Florence was well aware that the typical toothbrush does not clean either between teeth or along the gum line. Moreover, it is of little service to anyone with sensitive gums or gum disease, conditions known to be associated with ailments such as diabetes. Max knew that other brushing implements have been in use for millennia, including neem-tree twigs in India and miswaks through the Horn of Africa. By chewing neem sticks and miswaks, people fashion fine-pointed bristle brushes that clean between teeth better, lower bacteria, and reduce bleeding. But these chewing sticks have one common limitation; with bristles protruding straight from the stick, it is still impossible to clean behind back teeth. So in 1985 Dr. Florence designed a new hybrid brush with nylon bristles mounted at angles on both ends. This simple innovation makes it possible for anyone to stop bleeding gums within two weeks. More than thirty years later, Max's Canadian Sulcabrush is the better brush routinely recommended by dentists and periodontists. Worth a smile.

XYLANASE
The view from the bleachers.

For decades, the pulp and paper industry was the economic lifeblood of many cities and towns across Canada. Yet this same industry was also contaminating them. The use of chlorine, a vital ingredient in traditional pulp bleaching, led to vast amounts of toxic liquid waste being discharged into local waterways. In 1990, Canadian researchers developed an industrial enzyme that not only reduces the polluting liquid emitted from mills but also lowers the cost of producing pulp. These researchers manipulated the xylanase molecule so that it works at the high temperatures and acidity levels of pulp processing. The durable, practical enzyme is now used widely throughout North America and sold around the world, reducing chlorine pollution and production costs. It's even being explored as a way to make bio-fuels and lessen the demand for energy from petroleum products. The view from this bleacher is bright.

RAPID HIV TEST
The two-minute check-up.

Since he was a boy in Uganda, Abdullah Kirumira dreamed of becoming a doctor. He chased that dream for years, fleeing the violence of his birthplace first to Iraq and then Australia. It wasn't until he reached Canada — Nova Scotia, to be exact — that his dream came true. And more. As a professor of biochemistry at Acadia University, Dr. Kirumira became the driving force behind the province's biotechnology sector. Introduced in 1993, his greatest achievement is the world's first rapid-acting HIV diagnostic test. Performed in two minutes, it has revolutionized worldwide testing procedures for the virus by making tests quicker, cheaper, and more accessible than previous methods, which took up to five days to furnish their results. A quick, cheap, accessible test is a boon to care and prevention, giving countless patients knowledge of their condition and, if necessary, empowering them with the awareness they need to change their behaviour and begin treatment. One dreamer's two-minute test has changed the world.

GROWING NAIL
The implant that grows with the child.

Young children aren't small adults. They grow.
Seems obvious. Yet for years the only implant-
able devices that orthopaedic surgeons had at
their disposal to treat youngsters were those
designed for adults. Until 2000. That year, a team
of academics, researchers, and entrepreneurs in
the fields of orthopaedics, biomechanics, and
aerospace led by François Fassier and Pierre
Duval created the growing nail. The telescopic
implant is specially designed for very young
children who suffer from brittle-bone disease.
The growing nail does just that – its self-extending
rod grows, so to speak, with a child's femur,
tibia, and humerus even as it supports each of
these bones so they can grow at all. Because the
nail grows, it reduces the number of surgeries
that children afflicted with this disease must
undergo. It also puts much less stress on growing
bones than rigid rods, making growing pains a
thing of the past for, so far, some seven thousand
youngsters around the world. The admiration of
the medical community for this Canadian
innovation is evident in the growing nail's
official name – the Fassier-Duval.

CLIMATE RIGHTS
The one-woman paradigm shift.

Do universal human rights include the right to
be cold? Sheila Watt-Cloutier first thought so.
And now the world does. In 2005, she served a
landmark legal petition against the United States
that linked the effects of climate change on the
Arctic to the human rights of her fellow Inuit.
Nowhere is the influence of our warming climate
more apparent than in the Arctic. Ms. Watt-
Cloutier, who was born and raised in the Nunavik
region of northern Quebec, viewed and experienced the changes first-hand. The legal petition
was her way to start transforming her personal
experience into public policy and, in doing so,
change how the world views climate change –
not as an abstract or scientific phenomenon but
as an actual event that is upending the ways in
which people have lived and cultures have
endured for thousands of years. These people
have rights, including the right to be cold.

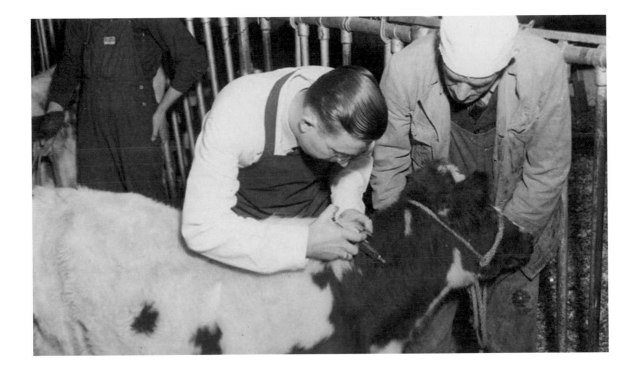

CATTLE PLAGUE VACCINE
The cure from Grosse-Île.

The first animal disease ever to be eradicated globally was Rinderpest. The virus, which strikes cows, sheep, pigs, and other split-hoof animals, spreads fast and has wiped out millions of livestock and wild animals over the centuries. No more. Credit goes to a group of Canadian scientists and veterinarians working in Grosse-Île, Quebec, who used chick embryos to develop the first Rinderpest vaccine for rapid mass production. The United Nations delivered large quantities of the vaccine to fight the disease in China and throughout Africa, and by 2011 Rinderpest was stamped out. The locale of this work is noteworthy. Located in the Gulf of St. Lawrence just north of Quebec City, Grosse-Île was the site of an outbreak of typhus in 1847 that killed thousands of Irish immigrants. That a vaccine to eliminate one disease emerged from a place associated with such tragedy is the most pleasing of ironies.

WRITE A BUSINESS PLAN: HERE'S HOW.

☐ Be aware that your business plan is the tool investors and partners will want to study before they decide to get involved.

☐ Think of your plan as your most current, most honest view of how your business will operate and succeed. Update it regularly. Refer to it often to see if you've hit the milestones you said you would.

☐ Create an executive summary (perhaps as a digital presentation) to tell your story briefly.

☐ Practise presenting the contents of your plan as a story about your business and your vision. It is through this pitch that investors will be able to grasp your idea, your product, your market opportunity, and your plan of execution.

WOUND DIAGNOSTICS
The clues under the skin.

A serious wound's ability to heal is determined by the level of oxygen-rich blood that reaches the wound. The method used to measure that level is surprisingly rudimentary: the naked eye. Physicians make their determinations of a wound's health based on its appearance on the skin's surface. Until now, that is. In 2012, Don Chapman of Kent Imaging in Calgary created something he calls the Tissue Viability System. Combining a camera and computer, the system targets the wound, calculates the amount of blood reaching the affected tissue, and produces a diagnostic image that gauges the ability of the wound to heal. The system does so by quickly snapping a series of photos using flashes of different light wavelengths that are safe for skin and eyes. The images are then blended to render a clear indication of the level of oxygenated blood reaching the wound bed and surrounding tissue. Although the system was designed initially to care for burn victims, it's now also being used to treat patients who suffer from diabetes, obesity, and other conditions that lead to poor circulation – proving genuine health is always much more than skin deep.

ITCLAMP
The haemorrhage control.

Bleeding to death in trauma is the leading cause of preventable death. Stop the bleeding, save the patient. The iTClamp does just that. Developed by Dennis Filips in 2013, the device goes over a wound and is squeezed together. That's it. The clamp, which anyone can use, takes three seconds to apply and shuts a wound tight. The Edmonton, Alberta, trauma specialist and retired Canadian Forces surgeon came up with the idea for the iTClamp while training medics on the best ways to stop bleeding while treating wounded personnel in the field. The disposable, inexpensive device isn't meant to be a permanent fix – just enough to stop the bleeding, limit blood loss, ensure patients don't require massive blood transfusions, and, above all, prevent them from dying before they reach the care of a trauma surgeon. Stop the bleeding, save the patient.

TELESURGERY
The remote operation.

Innovation often involves marrying seemingly unrelated methods to tackle pressing problems. In 2013, surgeons at the University Health Network in Toronto – led by Dr. Allan Okrainec – and engineers at Canada's National Research Council – led by Nushi Choudhury – brought together the latest advances in communications and simulation technology to provide long-distance teaching to neurosurgeons in Ghana. The need for teaching is plain. More than fourteen thousand young children are afflicted with hydrocephalus each year. Hydrocephalus occurs when fluid builds up inside a baby's skull and causes its brain to swell. Although the condition is treatable with surgery, it kills or impairs development when left untreated, which often happens because very few African surgeons are able to carry out the necessary surgery. The NeuroTouch virtual neurological simulator combines the latest surgical simulation tool with Skype and enables neurosurgeons in Toronto to train surgeons in Ghana to perform endoscopic third ventriculostomy, the delicate procedure required to treat hydrocephalus. The Ghanaian surgeons are now not only able to perform the vital surgery but can also become trainers themselves for other local surgeons, multiplying the impact of the initial training and extending a great marriage of technologies even further.

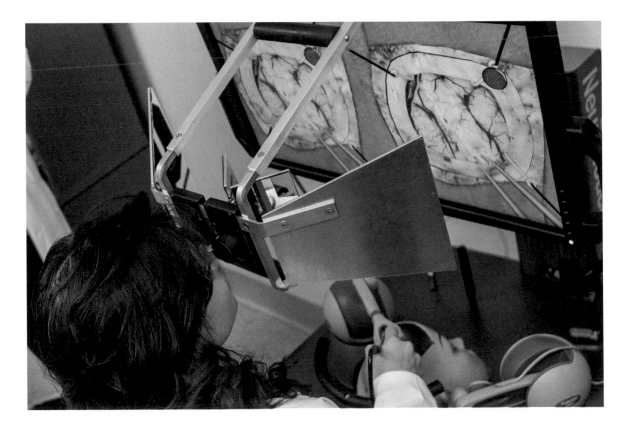

Pulp and Paper
Oil Drilling
Jerker Line
Chicken Bone
Syndicated Journalism
Marquis Wheat
Interest Calculator
Brownie Camera
Commercial
Canada Dry
Movie Theatre
Russel Logging Boat
Retail Cosmetics
Caisse Populaire
Superstardom
Jolly Jumper
Crispy Crunch
Chocolate Bar
Variable-Pitch Propeller
Rod Weeder
Whoopee Cushion
Plexiglas
Self-Propelled Combine Harvester

Coffee Crisp
Shreddies
Alkaline Battery
Instant Replay
Multiplex Cinema
Scarborough Suitcase
Instant Mashed Potatoes
Multi-dynamic Image
Bloody Caesar
Digital Photography
Air Seeder
Key Frame Animation
IMAX
Laser Dinghy
Saturday Night Live
Yuk Yuk's
Tutti-Frutti Modelling Dough
Trivial Pursuit
Loonie
Colour Coins
Ice Cider
Second Cup
Intelligent Teapot

Smarter
Smaller
Kinder
Safer
Healthier
Wealthier
Happier

While all societies depend on the livelihood of businesses for their prosperity, innovation is the single common force that makes those businesses successful. Canadian ideas have transformed countless industries—forestry, energy, entertainment, agriculture, finance, cosmetics, sports, film and television, transportation, and communications among them. And in doing so, Canada has proved that innovation itself is the new wealth of nations.

PULP AND PAPER
The better newsprint.

A better way is often found by paying attention to the goings-on of the natural world. Charles Fenerty could attest to the wisdom of this approach. One day in 1844, the Sackville, Nova Scotia, inventor watched several wasps chew wood fibres, transforming the pulpy chunks into papery strips, which they then used to make their nest. The wasps' work inspired Charles to develop the process of making paper from ground wood pulp – a much better process than the time-worn one of using rags. Yes, rags. He shared his new process with the owners of his local newspaper, the *Acadian Recorder*, and soon paper mills throughout North America adopted his method. Over the decades to come, these mills, fuelled by trees from Canada, churned out seemingly endless supplies of newsprint, giving rise to newspaper empires on this continent and in Europe. Built on an idea borrowed from a wasp, the information age really did get its start by going from rags to riches.

OIL DRILLING
The birth of an era.

Your friends in Texas might not know it, but the world's first oil well was dug in the Canadian town of Bothwell in 1857. That's right, the age of oil was born on the banks of the Thames River near a small hamlet in southwestern Ontario. The man behind the big dig was James Miller Williams. An innovative entrepreneur from Hamilton, he drilled that first hole to a depth of 27 feet, when it began to fill with a mix of oil and water. Abandoning the well when he could go no deeper, he started a new one in nearby Enniskillen. There he dug a well to 65 feet and began to draw from it between 5 and 100 barrels of oil each day – enough that Mr. Williams built the first oil refinery in Canada later that year to process what his well produced. The birth of the age of oil had begun. In Ontario.

JERKER LINE
The efficient pumper.

What do you do if it costs too much to operate a bunch of engines to power each of your dozen or so oil pumps? If you're John Henry Fairbank, you rig up a system of wooden rods to transfer the power generated by one engine to all of your pumps. In 1863, the inventive Oil Springs, Ontario, oilman built just such a jerker line. His wooden pump jacks were all connected to a series of wooden rods that, powered by a single engine, swung back and forth from wooden hangers. The hangers were suspended from wooden poles. You can still see it today. Swing on by.

CHICKEN BONES
The holiday confection.

What's so special about a chicken bone? No, not just any chicken bone. The Ganong Original Chicken Bone. Developed in 1885 by Ganong candy maker Frank Sparhawk, the Canadian confectioner's creation is a cinnamon hard candy jacket with a bittersweet chocolate centre. It made their candy a neat yet still sweet treat. Christmas would never be the same again.

SYNDICATED JOURNALISM
The insight readers wait for.

Anyone who grew up with Ann Landers and
Dear Abby may be forgiven for thinking that
advice to the lovelorn is an American innovation, but they'd be wrong. The honour must be
given (among many others) to Toronto's Kit
Coleman, who in 1889 began penning her
regular column "Woman's Kingdom" in the
Toronto Mail. While the bylined feature was styled
"for women only," it was avidly read by men too,
notably Prime Minister Wilfrid Laurier, who, by
his own admission, would never miss it.
Immediately celebrated for her frank advice and
uncommon insights into the complexity of
human personality, Kit was later dubbed "Queen
of Hearts." Kathleen Willis Watkins Coleman
was a relentless innovator and pioneer. She was
the first woman to be a page editor of a newspaper,
the first accredited woman war correspondent
when she covered the 1898 Spanish-American
War in Cuba, first president of the Canadian
Women's Press Club, and, with her internationally read "Kit's Column," the first syndicated
journalist in the world.

MARQUIS WHEAT
The grain in the world's breadbasket.

The growing season in Canada is notoriously
short. The time from the spring runoff in April
to the shortening daylight hours in September
seems like the blink of an eye. For wheat
farmers, the time must seem shorter still.
Thankfully, they have a strain of wheat designed
for Canadian conditions. Its name is Marquis
Wheat. Its creator is Charles Saunders. A cereal
specialist at Canada's Central Experimental
Farm in Ottawa, Saunders applied both scientific methods (such as cross-breeding wheat
varieties with desirable qualities) and rudimentary methods (such as chewing on kernels). The
result was a strain of wheat that grew fast,
produced high yields, and, because it was hard,
made flour that produced superior baked goods.
By 1891, he felt confident enough in the qualities of his new strain to send it to Saskatchewan
to be tested. Marquis passed: it matured fully a
week or two earlier than the dominant wheat
strain at the time, generated a bumper crop,
and made for a delicious slice of bread. Its
timing couldn't have been better. Waves of
immigrants were beginning to settle the
Canadian West. Marquis Wheat became the
staple crop as these new farmers turned their
land into the breadbasket of the world.

INTEREST CALCULATOR
The computerized tomato can.

Carle Meilicke ran a lumberyard in Dundurn, Saskatchewan. He spent the better part of his days calculating prices for various lengths and widths of wood, at least until one day in 1896. That's when Carle started experimenting with tables of calculations to price different lengths of wood. The result? The Meilicke calculator. His earliest version was a set of old tomato cans on which he etched scales of detailed information. Carle could rotate the drums to calculate answers to specific mathematical problems. Later versions were notebooks with tabs running along the tops. Finding a solution involved matching information on a tab with the right place on the table — a kind of early spreadsheet wrapped around a drum. Each calculator was created to solve one kind of problem — interest rates or personal income taxes or a company's payroll, but it was the interest calculator that became the big seller. Carle opened a factory in Chicago to build his custom calculators, eventually employing as many as forty people. The company that started in a Saskatchewan lumberyard in the nineteenth century was hard at work right up until the 1970s. It took seven decades for the computer to catch up with Carle's tomato can.

BROWNIE CAMERA
The magic name.

Until 1900, cameras were big, bulky, expensive contraptions, by their nature the exclusive province of professional photographers. At the beginning of the twentieth century, Eastman Kodak ended that era with the first in a line of inexpensive cameras anyone could operate. Along with its low cost and introduction of the snapshot, a major factor in the new camera's success was simply its name. The Brownie took its moniker from the fairy-like characters in a series of internationally popular books and comic strips by Palmer Cox, a Canadian writer and illustrator from Granby, Quebec. In Cox's illustrated world, brownies were kind-hearted sprites whose pranks made life more fun. Simple. Happy. Entertaining. Three ideal brand values for the new consumer gadget. Eastman Kodak approached Cox and a licensing deal was struck. More than 150,000 Brownie cameras were sold in their first year of production, many to children, proving that not only the right product but also the right name can work magic.

COMMERCIAL
The promotional spin.

Almost from the outset, motion pictures were used to sell as well as entertain. It all started in Montreal in 1903 when Sir Clifford Sifton persuaded executives of Canadian Pacific Railway to put up the money for an advertising film. The country's immigration minister hoped to attract settlers to the vast and largely empty expanse of Canada's Prairies. The railway execs wanted something too: paying customers. The railroad was the only way west for passengers from Canada's eastern seaports and the only way east for the grain these new farmers would soon be growing. The resulting forty-five-minute commercial – much of it shot in Canada's West – was wildly successful, generating a rush of European and Russian immigrants that continued for years. This new form of advertising also continued for years, all from a promotional spin on a new technology that made the world's first commercial film.

CANADA DRY
The champagne of ginger ales.

John McLaughlin was a successful pharmacist and entrepreneur. He was also a man of moderation with a private mission in life: to create a non-alcoholic alternative to champagne. He worked on his beverage for years. In 1904, it was ready: Canada Dry Pale Ginger Ale. The *Toronto Star* was impressed: "This ale has a mild yet piquant flavour, which is most agreeable, and its stimulating effect on the digestive organs makes it a prime favourite." Maude McLaughlin described her husband's new beverage more concisely as "the champagne of ginger ales." Its dry taste a refreshing change from the sweeter beverages that dominated the market, Canada Dry proved an immediate success. Two decades later, demand for the drink rose sharply during Prohibition in Canada and the United States. By 1930, more than ten million bottles a month were shipped to quench thirst across the continent. Yet while its tagline promoted Canada Dry as an alternative to alcohol, many employed Mr. McLaughlin's beverage as the ideal accompaniment to a shot of rye whisky. Many still do. Bottoms up.

MOVIE THEATRE
The magic house.

Few experiences are more magical than settling into your comfy seat as the house lights go down and the images begin to flicker and the sounds start to swell from the giant screen before you. You can thank Léo-Ernest Ouimet for the magic of the movie theatre. Before the Montrealer opened his Ouimetoscope on Saint Catherine Street in 1906, watching a motion picture was something done by small handfuls of people sitting on rickety chairs in any old room. Ouimet's motion picture theatre made going to the movies more like an evening at the *legitimate* theatre — formal, luxurious, and immersive. A true event. So say a quick thank you to Léo and the Ouimetoscope the next time you're about to enjoy the magic of a movie theatre.

RUSSEL LOGGING BOAT
The lumberjack's truck.

To the men who logged Canada's forests, rivers were highways that transported their harvest to the mills. And just as trucks ply the asphalt highways, Canadian lumbermen had trucks of their own. The first lumberjack's truck was the Russel logging boat. Built in 1907 by Russel Brothers Limited of Owen Sound, Ontario, the safe, fast, tough-as-nails boat was essential equipment on rivers throughout the Canadian Shield of Ontario and Quebec — the country's logging country. Steel-hulled and equipped with a power motor and winch, the boats were used to assemble and move large log booms down-river. They were also nimble enough to round up stray logs and herd booms over spillways and dams. The company made more than 1,200 of its distinctive logging boats. It also manufactured tugs and switching locomotives, and supplied vessels used in the Normandy landings on D-Day, boats that look quite similar to their logging cousins — essential equipment for both logging and liberation. So iconic was the Russel boat that, for many decades, most Canadians kept an engraving of one right in their pockets and purses . . . on the back of every Canadian one-dollar bill.

RETAIL COSMETICS
The new way to be ladylike.

Born in 1878, Florence Nightingale Graham learned about business at her father's side whenever they rode in their horse-drawn vegetable cart from Woodbridge, Ontario, to Toronto's St. Lawrence Market. Life was tough; the market trade earned little and, after her mother died young, Florence routinely went to bed hungry, shivering in the cold. She vowed to reverse her fortunes later in life. At the turn of the century, she made her way to New York and worked as one of the first treatment girls in one of the first beauty salons. Till then, makeup had long been considered a habit of the poor, especially of prostitutes. Attitudes were changing, however, and Florence was determined to lead the revolution. Foreseeing a time when women would have to wear makeup to be thought of as ladylike, Florence changed her name to the more dignified "Elizabeth Arden," opened her famous Red Door salon in Manhattan, and in time convinced a generation that the application of scientifically formulated colouring to eyes, lips, and skin was a reliable path to social acceptance. Elizabeth Arden led the creation of the global cosmetics industry and in the bargain became one of the richest women in the world. A long way from Woodbridge, Ontario.

CAISSE POPULAIRE
The people's bank.

One thin dime was the deposit required to become a member of Alphonse Desjardins's Caisse Populaire de Lévis when it opened in 1901. The Quebec reporter found inspiration for his unconventional financial institution in two sources. The first was the concept of credit unions, which had sprung up in Europe a few decades earlier to provide working people with loans at fair rates of interest. The other influence was the account of a Montreal man who had been ordered by a court to pay a moneylender some $5,000 in interest on a $150 loan. Yet Desjardins's bank of the people was a little different than its European cousins: along with providing credit at reasonable rates, it stressed inclusiveness (all it took to be a member was that ten-cent deposit), community development (these local financial institutions played roles nurturing the places in which they operated), and faith (they were organized according to parishes that divided administration of the Catholic church in the region). This better model for the credit union soon spread throughout the province, into the northeast United States, and eventually throughout Latin America and South America. Quebec's people's bank is now the world's.

SUPERSTARDOM
The dawn of Hollywood celebrity.

Hard to imagine that it took a sickly Canadian girl to turn Hollywood stardom into a full-blown cult. Even though born into an impoverished Toronto family in 1893, and despite a childhood

marked by diphtheria, pneumonia, tuberculosis, and a variety of nervous disorders, Gladys Louise Smith was destined to become the first globally recognized movie superstar. Supporting her family as a child actor by taking small-stage roles in Ontario, she moved to New York in 1909 to try her luck with films using a stage name she had coined just two years prior – Mary Pickford. Admired immediately by filmmaker D.W. Griffith, she was hired for twice the going rate at $40 per day and appeared in fifty-one films that first year. By 1916, her weekly salary was $10,000 – the highest fees ever paid to any actor, and by newspaper accounts the most paid to any woman in the world in any field. Everywhere she went she was met with throngs of fans hoping for a smile or an autograph. A dogged innovator by nature, Mary Pickford created United Artists with her colleagues Griffith, Charlie Chaplin, and Douglas Fairbanks in 1919, and in 1927 was instrumental in forming the Academy of Motion Picture Arts and Sciences, which two years later voted her the Oscar for Best Actress. Though she remained a resident of California, she insisted on keeping her Canadian citizenship, supported the war effort by investing heavily in Canadian War Bonds, and of course dined with Canadian luminaries such as Prime Minister William Lyon Mackenzie King when back on home turf.

JOLLY JUMPER
The back saver.

Life is hard enough with two arms. When one of them must hold a squirming youngster, it can be downright impossible. After her first child was born in 1910, Toronto mother Susan Olivia Poole was keen to stay active. Inspired by the papooses used by Aboriginal mothers to carry their children, she fashioned a harness of her own. It was a cotton diaper fashioned as a sling seat, a coiled spring to suspend its wearer from above, and an axe handle to secure the contraption. Susan called her combination a Jolly Jumper. As she worked in home and garden, her son bounced playfully and safely nearby in his new jumper, toes just off the ground. Susan had six more children and made Jolly Jumpers for each. When her children had children of their own, Grandma Susan made even more. In 1948, she began to build and sell them, eventually patenting her jumpers and selling them farther afield. Today, Susan's jolly little harnesses are no longer made out of diapers, springs, and axe handles, but they are all hard at work freeing the arms and saving the backs of countless grateful parents around the world.

CRISPY CRUNCH
The $5 winner.

Peanut butter and chocolate seem like a natural combination. Harold Oswin sure thought so. In 1912, the candy maker at Toronto's William Neilson Dairy entered his idea of a chocolate bar with a peanut butter centre in the company's contest to find a new bar. He won, pocketing five dollars in the process. For the next seventy-five years, the Crispy Crunch just muddled along—never a favourite, but never so unpopular as to be dropped from the chocolate-maker's roster. Then in 1988, a successful brand repositioning launched Crispy Crunch straight to the top of the chocolate-bar heap. The campaign's catchphrase: "The only thing better than your Crispy Crunch is someone else's." On such few words rests the fate of chocolate-bar empires. The winner of the five-dollar prize was finally number one.

CHOCOLATE BAR
The confection that launched an industry.

Take a good idea, wrap it up, and put your name on it. That's how the multi-billion-dollar chocolate-bar industry was born. It was the summer of 1920. New Brunswickers Arthur Ganong (president of his family's chocolate factory) and George Ensor (its superintendent) took an afternoon to go fishing. As they gathered their gear, they slipped chunks of chocolate in their pockets to enjoy on their outing. The treat satisfied their sweet teeth but left them with messy hands and messier pockets. So next time out they wrapped their snacks in Cellophane, a new kind of material that repelled liquids (including molten chocolate). That's when the light went on. If they could enjoy such convenience, why couldn't everyone? The Ganong company began to manufacture and sell a two-piece chocolate bar enclosed in a sharp-looking red-and-yellow Cellophane wrapper. They called the chocolate, fudge, and peanut combination Pal-o-Mine – a friend for fishing trips and any other moment in life that might be sweeter with a snack.

LAUNCH A START-UP: HERE'S HOW.

☐ Find one or two co-founders with complementary skills who agree with your business vision and who would like to work with you for a long time.

☐ Sort out the split of company ownership early on.

☐ Incorporate and put some money into the company.

☐ Ask for help with your start-up from your government's Industrial Research Assistance Program or equivalent. Find out what that group can do for you, and what it will require of you in return.

☐ Build your prototype based on customer discovery.

☐ Show friends and family your prototype and see if they will contribute some financing.

☐ Find a cost-effective place to work with your team – either at home or at a business accelerator. Save your money for building your product and validating your market opportunity with early customers.

VARIABLE-PITCH PROPELLER
The dawn of air transport.

It's the transmission for a plane. Invented in 1928 by Wallace Turnbull of Rothesay, New Brunswick, the variable-pitch propeller is designed to move an aircraft in different directions and adapt to different conditions. Pitch is the angle of each blade of the prop. Low pitch enables a plane to climb more effectively. High pitch optimizes performance and economy at high speeds. Blades can also be pitched to create a braking effect and enable planes to land in shortened distances. Before being equipped with variable-pitch propellers, aircrafts operated in one gear at all times. These one-trick ponies had difficulty getting off the ground with large payloads and couldn't travel great distances. Armed with one, the sky's the limit. Wallace's ingenious prop is quite simply one of the most vital advancements in the history of transportation.

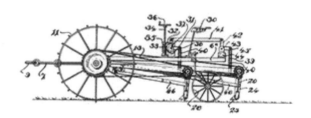

ROD WEEDER
The gentle field treatment.

Innovation is sometimes just doing the obvious thing at the right time. That's how modest George Morris characterized the device he developed. In 1929, the Bangor, Saskatchewan, garage owner, tractor dealer, and machinist created a new kind of rod weeder. Conventional weeders were horse-drawn machines with a cylindrical steel rod that dragged through the soil, rotating and removing weeds as they went. In the province's rocky soil, these tools were often bent out of shape or even snapped off entirely. Compounding the problem, traditional weeders dug deep, loosening much of the rich topsoil and exposing it to the stiff Prairie wind. George's new rod weeder solved both problems. It featured an automatic mechanism that tripped the rod when it came into contact with rocks. The trip mechanism would spring the rod to the surface of the field to avoid the potentially damaging obstacle and then snap it back down to continue its work. The sensitivity with which the device operated also left the soil cover intact. As the Great Depression hit and the winds of the Dust Bowl began to blow, George's new rod weeder was the right thing at the right time.

WHOOPEE CUSHION
The new sound of novelty.

A new sound: that's all a novelty item needed to become a raging sensation in the late 1920s. Companies offered a wide variety of devices that emitted strange sounds when squeezed — some a child's scream, others a cat's screech. Experimenting with sheets of rubber, employees of the JEM Rubber Company in Toronto hit upon a diffcrent sound. The noise that emanated from their little rubber pillow was a tad more, how shall we put it, indelicate. American novelty purveyor Johnson Smith & Company heard the call and added JEM's doohickey to its giant catalogue. The economy model went for 25 cents, a deluxe edition for $1.25. A perfect gift for the discerning prankster who has everything. Sales erupted with a loud toot and haven't ceased. The sound of the Whoopee Cushion can still be heard loud and clear wherever unsuspecting bottoms and chairs get together.

PLEXIGLAS
The safe windowpane.

Just who invented Plexiglas? The exact origins of the safe windowpane are ironically unclear. Some say the development of transparent polymerized methyl methacrylate dates to 1877. The Germans made this acrylic substance back then but couldn't produce a version that wasn't cloudy. Plexiglas must be clear to be of much use, mustn't it? That's what William Chalmers thought. As a graduate student at Montreal's McGill University in 1931, William decided to bring some clarity to the subject. He discovered that methacrylic ethyl ester and methyacrilic nitrile could be joined at the molecular level to create a transparent polymer. He patented this technique and later sold it. Two years later, it appeared on the market under the proprietary name Plexiglas, owned by another German. Variously referred to as Lucite, Acrylite, Perspex, and Plexiglas, this ubiquitous feature of modern life makes our view safe and clear. German invention, Canadian refinement. Innovation at its collaborative best.

SELF-PROPELLED
COMBINE HARVESTER
The do-it-all machine.

Modern harvesting began in Canada. In 1937, Massey-Harris introduced the world's first self-propelled combine harvester. The new machine was the brainchild of Tom Carroll. Tom's harvester did exactly what it advertised. It harvested crops. It was a combine that performed or "combined" all the necessary elements of the job — reaping, threshing, and winnowing — in a single process. And, most originally, it was self-propelled: the Massey-Harris machine featured a harvest header placed up front so as not to damage crops with its wheels; the driving position of the harvester was above the centre of the header, giving the operator an excellent view of the cutting action. Steering wheels at the back of the combine also made the new machine more manoeuvrable than a combine drawn by a tractor. Self-propelled combine harvesters have been refined since to make them more efficient. But their basic designs hold true to the Canadian do-it-all original.

SHREDDIES
The good, good whole wheat.

What's so special about Shreddies? First manufactured by Nabisco in Niagara Falls, Ontario, in 1939, Shreddies was the first brand of cereal made from strands of whole-grain wheat interwoven like yarn in a sweater. While only mildly popular in its home country when first launched, Shreddies were exported to the United Kingdom in 1955 and soon became a breakfast favourite. Of course, you'll find the iconic cereal brand on grocery shelves in both countries today. So eat up, everyone, they'll knit more.

COFFEE CRISP
The nice light Canadian.

Would you like coffee with your chocolate bar? Rowntree did. In 1939, the Toronto confectioner added a creamy coffee filling to the layered chocolate wafer of a British bar known as Chocolate Crisp. Voilà, Coffee Crisp. It remains a favourite, but sadly mostly in Canada. Canadians abroad have long lamented that the nice light snack is largely unavailable elsewhere on the planet. A nice, light, singularly Canadian snack—that's what you are.

SCARBOROUGH SUITCASE
The retractable handle.

Carrying a twelve-pack of your favourite brew home from the store was once a two-handed task. Then Steve Pasjack brought civilized behaviour to the act. In 1957, this enterprising beer drinker created the first retractable handle for a beer carton. It slides straight up outside the case when you want to carry it, then slips out of sight when it comes time to store the case. Case and handle together were nicknamed the Scarborough Suitcase. Was Steve even from the Toronto suburb? Doesn't matter. The distinctive handle lasted for decades until replaced by a variety of less imaginatively named options . . . but not completely. Some breweries in Toronto still feature the classic handle on their cases. Try one, and have one for Steve while you're at it.

INSTANT REPLAY
The story worth repeating.

George Retzlaff is the most influential person in the history of sports you've never heard of. As producer of *Hockey Night in Canada*, he came up with instant replay in 1955 and made it a regular feature on CBC Television's flagship broadcast. The advent of videotape the following year made the process much easier, and therefore instant replays became an increasingly frequent and imaginative element of broadcasts. Big deal, you say? While a television broadcast of a game prior to Retzlaff's brainchild was merely a substitute for attending the real thing, instant replay made a broadcast a viewing experience unto itself. It changed not only how we watch sports but also how the games themselves are played, analyzed, and even officiated. A case of the medium influencing the action. Now let's take a look at that again: George Retzlaff is the most influential person in the history of sports you've never heard of. As producer of *Hockey Night in Canada*, he . . .

MULTIPLEX CINEMA
The dark choice.

If you walk down Elgin Street in Ottawa, you can still see the red brick facade of the world's first multiplex cinema. The man behind the screens is Nat Taylor. The owner of the Elgin Theatre, Nat opened his movie palace in 1937. Business was good, so he raised a second screen on an adjacent plot of land ten years later. By 1957, however, Nat had become frustrated at being compelled by powerful film distributors to stop showing one popular film on his screen just to replace it with another. Nat countered by keeping a popular old release on one screen while premiering a new release on the second. With this simple innovation, he created the multiplex. Nat went on to build larger multi-plexes in hundreds of suburban neighbourhoods across Canada. Other entrepreneurs built their own versions in neighbourhoods throughout the movie-loving world. And the Elgin Theatre? In 1994, the lights flickered off at Nat Taylor's modest two-screener. Once the dark choice of moviegoers, the first multiplex cinema had become a victim of its own success.

ALKALINE BATTERY
The fresh start.

Sometimes the best way to improve a device is to toss it aside and come up with something else entirely. When asked by his superiors at the Canadian National Carbon Company to improve the longevity of traditional zinc-carbon batteries, Lewis Urry quickly decided on a better approach. In 1955, the newly minted engineer from the University of Toronto experimented widely with other elements and compounds. He discovered that a combination of manganese dioxide, solid zinc, and powdered zinc would produce not only a long-lasting battery, but also one that was affordable and easy to produce. The secret ingredient, according to Lewis, was the powdered zinc, which gave the battery more surface area and therefore more power. Next time you pick up your smartphone or turn on your laptop, think of Lewis Urry and the benefit of starting fresh.

INSTANT MASHED POTATOES
The flakey idea.

The gallant gourmet's horrific nightmare is the harried homemaker's blissful dream. Edward Asselbergs developed a way to dehydrate the humble spud into flakes in 1960 while working as a chemist at Agriculture Canada in Ottawa. His instant mashed potatoes, a staple of so many modern meals, were found in grocery stores everywhere a mere two years later. A dream comes true, one box of flakes at a time.

MULTI-DYNAMIC IMAGE
The pictures in the picture.

If you watched television in the seventies, you were well aware of Christopher Chapman's brainchild – even if you didn't know what it was or who he was. His multi-dynamic image technique was a staple of the opening credits of several shows: *Kojak*, *Mannix*, *Barnaby Jones*, *Medical Center*, and many more. The technique – developed by the Torontonian in 1967 and first featured in his own film, *A Place to Stand* – involved multiple images shifting simultaneously on panes across the screen. Some panes were whole images moving about, while others contained parts of a bigger image that takes shape as the panes appear and coalesce like puzzle pieces. Remember the introduction to *The Brady Bunch*? This opening sequence is a perfect example of the technique in action, with as many as nine independent images on the screen at one time. Nine pictures within the picture – and Ann B. Davis as Alice.

BLOODY CAESAR
The Canadian cocktail.

Walter Chell was up for a challenge. In 1969, the bartender at Calgary's Westin Hotel was asked to create a signature drink to mark the opening of a local Italian eatery. He stewed on the task for three months, experimenting with various concoctions before coming up with a novelty that combined vodka, hand-mashed clams, tomato juice, Tabasco and Worcestershire sauces, salt, and pepper, all garnished with a celery stick. The mixologist called his libation a *Caesar*, a nod to his Italian roots, and not without a little swagger. Then one day, the story goes, a tart-tongued patron took a swig and declared, "Walter, that's a damn good bloody Caesar!" and the full name was born. He wasn't the only one who felt that way. The cocktail became popular so quickly that within that same year the beverage maker Mott's began selling record amounts of its ready-made mix of clam and tomato juices. No more mashing clams. Today, the company claims that Canadians hoist 350 million of these national cocktails every year. Hail Caesar!

DIGITAL PHOTOGRAPHY
The new way to say cheese.

"Come up with something new." One day in 1969, Willard Boyle received this instruction from his Bell Laboratories boss. The impetus to innovate can be just that blunt sometimes. So the Nova Scotia native got to work, and fast. During a single afternoon's work at the blackboard, he came up with the charged-couple device. What's that, you ask? It's the "something new" that lets us take photos with our smartphones. A charge-coupled device moves an electrical charge along the surface of a semiconductor from one storage capacitor to the next, allowing incoming photons to be converted into electrical charges, which can then light up a portion of a display. With enough capacitors, one can form a glowing image of the original view. Although Willard had the idea, it took a team of experts to make his vision a reality. The result? Innovation through inspiration and teamwork.

AIR SEEDER
The idea worth spreading.

For generations, farmers relied solely on gravity to seed and fertilize their fields. They filled large seed boxes attached to their farming implements. The machines then released seeds or fertilizer to fall into seed rows. Gravity is a powerful but blunt force. Farmers often had to make many passes through a field to seed and fertilize it properly. In 1970, Jerome Bechard devised a better way. The Lajord, Saskatchewan, farmer's machine features a tank pulled by a tractor. The tank is separated in compartments that carry seed and fertilizer. A stream of compressed air distributes a mix of seed and fertilizer directly and forces it deeply into seed rows. The tanks also have a meter that farmers manipulate to adjust the mixture of seed and fertilizer. Jerome's air seeder spread quickly. Instead of gravity, it is now the standard wherever grain is grown.

CONTROL THE USE OF YOUR IDEA: HERE'S HOW.

While innovation depends on widely shared ideas, many nations (including Canada) allow creative thinkers the exclusive commercial right to implement their ideas for specified ranges of time. Ideas protected in this way are referred to as intellectual property or IP. Here are steps you can take to control your own IP.

☐ **Create an IP strategy early on suited to the industrial sector you operate in, the type of IP you want to protect, and your financial capacity to defend that IP down the road.**

☐ **Put a solid technical writer on your team to draft provisional patents with one or more patent ideas each. Track which of those ideas are most valuable to your firm in the first year.**

☐ **Search for and respect other people's patents and barriers such as registered marks. This will avoid later legal interruptions.**

☐ **Have an IP lawyer help you create clear title to the ideas you want to control, with enforceable employment and contractor agreements that assign title to all IP, including patents, copyrights, trade secrets and moral works. Ensure that employees and contractors enter into and re-acknowledge those agreements every year.**

☐ **Get a corporate legal firm to help. Keep costs down by choosing a firm that offers an IP start-up package. Apply for patents, and also ask for help to protect your trade secrets and industrial design with confidentiality agreements and other tools.**

☐ **If someone claims you are infringing an existing patent, get legal assistance right away.**

KEY FRAME ANIMATION
The birth of computer graphics.

How do you create a $200-billion industry? Start with one challenge to overcome. In the 1960s, Montreal engineers Nestir Burtnyk and Marceli Wein wanted to make computers easier to use. When they heard animators at Disney explain the time-consuming process of making cartoons, the two found the challenge they needed. By 1970, Burtnyk and Wein had developed key frame animation. Their advance made it possible for an artist to sketch only the main points of an animated sequence; the computer program then calculated and produced the frames in between the main ones, eliminating the tedious manual work traditionally required as junior animators meticulously drew every cell, frame, and movement of a sequence. It did more: key frame animation paved the way for much more sophisticated computer-generated imagery—a $200-billion industry and a mainstay of film production today.

IMAX
The really big show.

Moviegoers want to be amazed continually by what they see, and moviemakers try to keep up and even get ahead by creating bigger, wider, clearer pictures. Today's truly big picture is IMAX. Before this system, filmmakers couldn't use the bigger film stock required to display images that would fill a viewer's field of vision. Seventy-millimetre film, which is ten times larger than regular film, would shake when run through a camera and projector, distorting the image these machines tried to capture and display. Created in 1971 by five men working in Toronto – Ron Jones, William Shaw, Roman Kroitor, Robert Kerr, and Graeme Ferguson – IMAX solved this problem by running the film sideways in a wave-like action called a rolling loop. This method produces a steadier (five times more than conventional systems), clearer (more than 17 million pixels per frame), and bigger image (ten times the size of commercial 35-millimetre film, or eight storeys high). Now that's seeing the big picture.

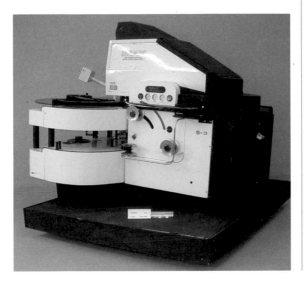

LASER DINGHY
The lake rocket.

One day in 1970, Bruce Kirby and Ian Bruce were talking over the phone about creating a line of camping equipment. They agreed that their all-Canadian gear must include a boat that weekenders could carry to the lake on the roof racks of their cars. Kirby got to work right away, sketching out a blueprint for their portable sailboat. His straightforward design emphasized simplicity, durability, and speed. When it was unveiled in 1971 at the New York Boat Show, the dinghy – now named Laser – was an immediate hit among boating buffs. Within three years, a world championship for the class was held. In 1996, the Laser became a men's Olympic-class boat at the Summer Olympics in Atlanta. And at last count, more than 200,000 Lasers were scooting about lakes and rivers in dozens of countries. The idea of two Canadian campers is now a mainstay for a world of weekend boaters.

SATURDAY NIGHT LIVE
The new comedy institution.

The brain behind the most enduring and influential weekly television comedy show in the United States is, funnily enough, a Canadian. Toronto's Lorne Michaels (pictured here) created *Saturday Night Live (SNL)* in 1975. First called *NBC's Saturday Night*, the program was meant to replace that night's rerun of a TV institution — *The Tonight Show Starring Johnny Carson*. Michaels's fledgling show quickly became an institution of its own. Other comedy programs had appeared on American TV for years, but *SNL* proved to be

something altogether different. It featured an ensemble cast of brash young comedians; it was broadcast live in front of a raucous studio audience; and it specialized in cutting social and political satire. It's never looked back. For nearly the past half century, *Saturday Night Live* has been copied many times but never duplicated. It has also served as a launching pad for many of the most beloved comedians of the past two generations, many of them Canadians like their show's creator. Funny that.

YUK YUK'S
The funny franchise.

If restaurants can be franchised, why can't comedy clubs? Mark Breslin asked himself that question in 1978, two years after the comedian and entrepreneur had opened his first comedy club in the basement of a Toronto community centre. He had called it Yuk Yuk's and sensed from its immediate success that a rapid stand-up roster model of entertainment could be popular anywhere. Soon after, franchises of his venue began popping up in cities across Canada. Today, there are seventeen Yuk Yuk's comedy clubs in Canada and three in the United States. Their success has also spawned imitators – many franchises of funny. Mark surely doesn't mind. The revolution in comedy begun by the Toronto entrepreneur has been a boon to comedians and stand-up fans throughout North America. Thanks for reading. You've been a great audience. Seriously. Good night.

TUTTI-FRUTTI MODELLING DOUGH
The toy you want to eat.

Never underestimate the willpower of a mother. In 1980, Micheline Desbiens tried more than five hundred recipes before she was happy with the one that became Tutti-Frutti Modelling Dough. Why so many takes? Unlike all the plasticines then on the market, this Quebec mother's play dough is non-toxic, rehydrates, doesn't stick or stain, has a pleasing odour, and can be twisted and pressed out of shape by small hands. What more could a mother want?

TRIVIAL PURSUIT
The uncommon phenomenon.

Here's a question from the Sports and Leisure category: Which Canadian-made game has been called the biggest phenomenon in board game history? Trivial Pursuit, the brainchild of four Montreal men, proves that entertaining a generation isn't complicated. All it takes is a board, six playing pieces, dozens of small coloured wedges, and six thousand questions divided into six categories on a thousand cards. Plenty of persistence also helps. Early investors were hard to come by, start-up costs were high, and initial sales were a bust. The good news: the first few people who played Trivial Pursuit were captivated and it soon became a board game sensation. Since its inauspicious release in 1982, more than 100 million copies of the game have been sold in 26 countries and at least 17 languages. Hardly trivial.

LOONIE
The eleven-sided token.

It might have ended up being called the Voyageur or maybe the *voygie*. In 1987, Canada's new one-dollar coin was ready to be struck with a scene depicting voyageurs paddling their canoe toward some unknown adventure. But the master dies to mint the coins were lost en route to the coin production facility in Winnipeg from Ottawa. The team at the Royal Canadian Mint turned to a back-up design, a simple image of a common loon created by Sault Ste. Marie native Robert Ralph Carmichael. Eighty million of the eleven-sided, bronze-plated-nickel dollar coins were circulated across Canada to replace the dollar bill. The reason for the switch is simple enough: a bill lasts no more than one year, while the new coins last at least twenty. A big factor in its resilience is a new electroplating process invented by Sherritt International, a multinational resource company based in Toronto, that enables the golden-hued coins to resist tarnish and wear. Within weeks of being released, the new coins were dubbed *loonies*. In 1996, their two-dollar cousins arrived on the scene. The *toonie*, what else?

COLOUR COINS
The noticeable change.

Bills of different colours have been around for many years. It wasn't until 2004 that the first coloured coins burst on the scene. In October of that year, the Royal Canadian Mint put 30 million of its red poppy 25-cent coins into circulation. The poppy was coloured with a high-speed, computer-controlled precision inkjet process. The coins, which were dedicated to Canadians who have died in service to their country, not only were legal tender but also became keepsakes for collectors. The Mint repeated the feat four years later to commemorate the ninetieth anniversary of the end of the First World War. It was a further burst of colour to mark the end of a dark chapter in our nation's life.

ICE CIDER
The pinnacle of applehood.

The strict definition of ice cider is a fermented beverage made from the frozen juice of apples. An equally accurate characterization is the pinnacle of applehood. Or is it appledom? Whatever the case, ice cider is the cider equivalent of ice wine. Invented by Christian Barthomeuf in 1990, ice cider is created in one of two ways. Cryoconcentration involves harvesting apples late in the season and then storing them until the end of December, when they are pressed and the juice left to freeze naturally and begin fermenting. Cryoextraction requires leaving the apples on their trees until late January. They are then picked, pressed, and left to cold-ferment. The birthplace of ice cider, Dunham in Quebec's Eastern Townships, is the ideal place for both these approaches — lots of apples and lots of cold late fall and winter weather. Some would say the town is itself the pinnacle of applehood. Or is that appledom?

SECOND CUP
The eye opener.

Back in 1975, Tom Culligan of Belledune, New Brunswick and Frank O'Dea of Montreal, Quebec discussed a growing trend — Canadians preferred coffee to tap water! They capitalized on the habit by establishing coffee kiosks conveniently located in local shopping malls — first in Toronto, then across Canada and around the world. Knowing that Canadians typically begin their days with coffee at home, they named their enterprise Second Cup, and the idea caught the public imagination. Second Cup Coffee Co. is now Canada's largest specialty coffee retailer, handcrafting over a million beverages and 50,000 pounds of fairly traded coffee every week. How do you take yours?

INTELLIGENT TEAPOT
The brew master.

How can a financial analyst spend his retirement years? One used them to invent the perfect teapot in 2009. Pierre Mercier is his name. The Montrealer's creation, the Fine T Machine, has two things going for it. It's programmable, so that tea drinkers can infuse their type of tea for exactly the right length of time in exactly the proper temperature of water. It's also made from stainless steel, so that tea drinkers can brew all kinds of teas in their pots without experiencing the tastes of different teas that build up over time in ceramic pots. And it's from Canada, you say? That's no pity.

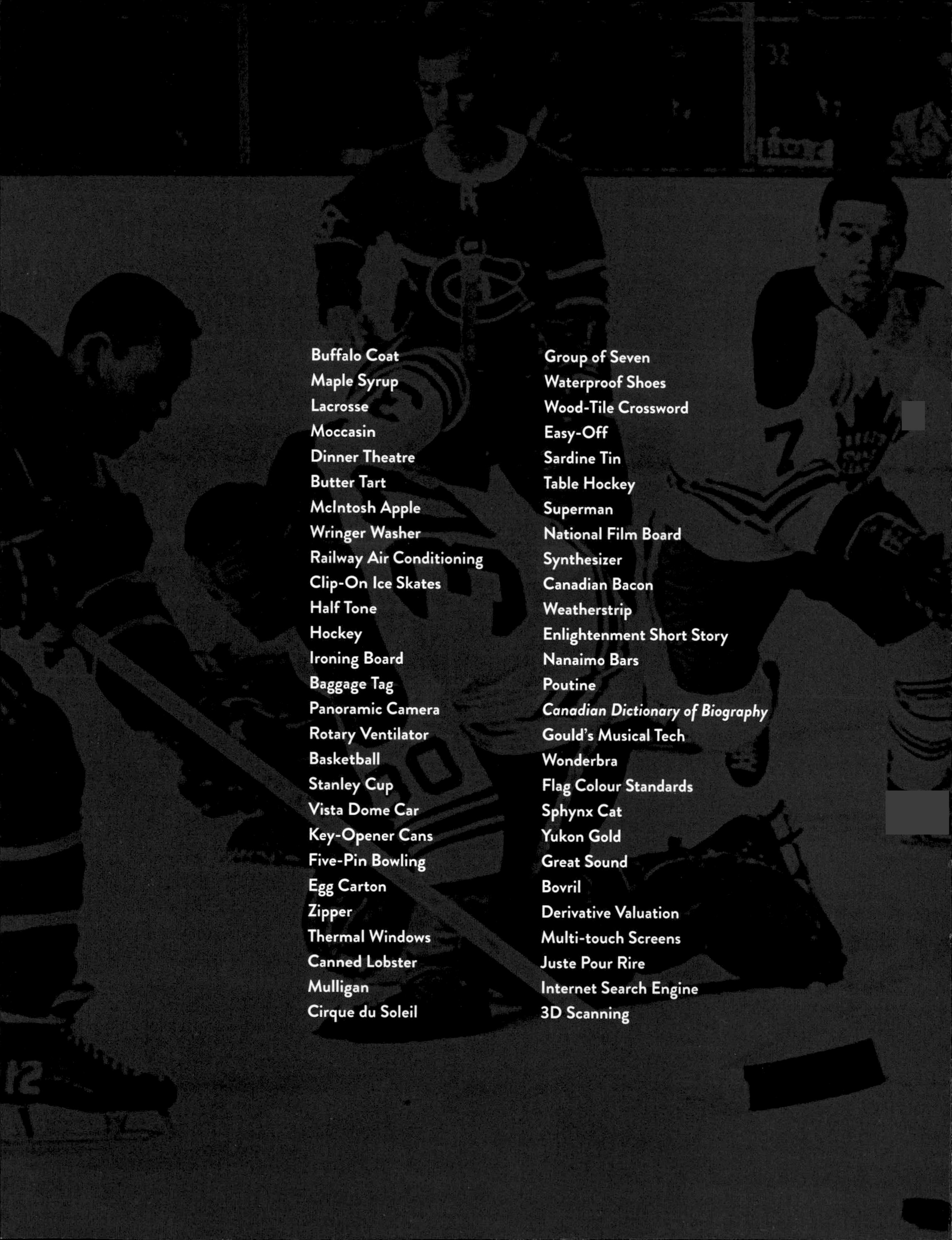

Buffalo Coat
Maple Syrup
Lacrosse
Moccasin
Dinner Theatre
Butter Tart
McIntosh Apple
Wringer Washer
Railway Air Conditioning
Clip-On Ice Skates
Half Tone
Hockey
Ironing Board
Baggage Tag
Panoramic Camera
Rotary Ventilator
Basketball
Stanley Cup
Vista Dome Car
Key-Opener Cans
Five-Pin Bowling
Egg Carton
Zipper
Thermal Windows
Canned Lobster
Mulligan
Cirque du Soleil

Group of Seven
Waterproof Shoes
Wood-Tile Crossword
Easy-Off
Sardine Tin
Table Hockey
Superman
National Film Board
Synthesizer
Canadian Bacon
Weatherstrip
Enlightenment Short Story
Nanaimo Bars
Poutine
Canadian Dictionary of Biography
Gould's Musical Tech
Wonderbra
Flag Colour Standards
Sphynx Cat
Yukon Gold
Great Sound
Bovril
Derivative Valuation
Multi-touch Screens
Juste Pour Rire
Internet Search Engine
3D Scanning

Smarter
Smaller
Kinder
Safer
Healthier
Wealthier
Happier

Some innovations make us feel immediately better. The comfort of a soft moccasin on the forest floor. The thrill of a Stanley Cup final game. The taste of Canadian bacon. The sound of a perfectly balanced speaker. The wit of a stand-up comic. The cool of an air conditioned vehicle. The taste of maple syrup. A second chance in golf. All of these owe their pleasing effects to the creativity of Canadians. If it weren't for these innovations, the world would be a sadder place.

BUFFALO COAT
The warmest winter wear.

People have tanned hides for almost as long as they have hunted. The Indigenous peoples of Canada were among the first tanners, with Cree, Assiniboine, and Blackfoot perfecting their art on the Prairies. They favoured the hides of bison, animals that roamed in the millions across their shared land. In the days of western settlement by Europeans, these hides were fashioned into coats with sleeves and buttons, commercially produced, and sold to groups operating in extreme cold. Notable among their customers were the Royal Canadian Mounted Police, who were often depicted in engravings, photographs, and eventually films in their iconic buffalo coats. When the Prairie bison were hunted to near extinction, the fashion changed in time to the lighter, simpler, and (especially if you're a bison) *kinder* parka.

MAPLE SYRUP
The rite of spring.

A stack of pancakes just wouldn't be the same without maple syrup. The Algonquin of eastern Canada were the first to recognize the dietary value of the sap from red, sugar, and black maples. For countless generations, these Indigenous peoples drank maple sap as a sweet drink and used it in cooking as a source of energy and nutrients. European settlers introduced the idea of using iron and copper kettles to boil sap and evaporate some of its water, transforming it into not only syrup but also maple sugar, which soon became a staple on the tables of colonial homes. Advances over the last several hundred years have ushered in tools and processes to quicken evaporation and improve syrup quality. Most recently, large-scale operations have incorporated vacuum pumps, pre-heaters, reverse-osmosis filters, and tube systems that connect trees to the evaporator house. The rite of spring is big business. Canada produces 80 per cent of the world's maple syrup and exports it to hungry diners around the world in all seasons.

LACROSSE
The sport that honours the Creator.

Imagine a sport in which two teams each of five hundred players compete all day on a field three kilometres in length. Indigenous peoples in Canada believed that lacrosse, whose origins have been traced back at least a millennium, was something to be played on a grand scale — one worthy of the Creator in whose honour the contest took place. When European settlers arrived in Canada, they wrote about these sacred spectacles of sport. In time, they participated. When the national conversation turned to Confederation in the late 1850s, William George Beers of Montreal set himself a project. William loved lacrosse. He had grown up playing it and believed that, properly promoted, the game could serve as a "unifying symbol for the emerging Canadian nationality." By 1860, he had codified the rules of the sport. Before the decade was out, he had written and published the world's first book on the game. In 1876, when the sport had caught on, he sailed with a team of Indigenous and non-Indigenous Canadian players to showcase lacrosse throughout England, Scotland, and Ireland. Today, this ancient Canadian sport is played by millions throughout North America and around the world, by men and women, professionals and amateurs, outdoors and indoors.

DINNER THEATRE
The order of good cheer.

MOCCASIN
The footwear of the forest.

Giving protection and connection, moccasins are the ideal footwear for forest living. Thousands of years old, their exact origin is unknown. But for many First Nations peoples in Canada, especially the Algonquin, they were the perfect boot. Made entirely of leather, most moccasins are made up of a sole of leather that has not been softened and sides of supple leather stitched together at the top so they enclose the foot fully. The sole is hard enough to protect the foot yet soft enough to enable the wearer to feel the ground. Upon their arrival in North America, European hunters, traders, and settlers saw the benefits of moccasins, quickly and gratefully adopting them as their own footwear of the forest.

Samuel de Champlain had a good reason to throw a party. The earliest French settlers in Canada believed idleness caused scurvy, a bout of which had just ravaged their fledgling community on Saint Croix Island on the Bay of Fundy. So many perished that they moved the colony across the bay onto the mainland of what today is Nova Scotia. In November 1606, as another merciless Canadian winter descended on Port-Royal, the governor of New France established the Order of Good Cheer. The eating and theatrical club involved plenty of homegrown food and entertainment to keep the settlement's seventy men active and the killer ailment at bay. Gatherings of the Order occurred weekly through the winter, ending in the last week of March and starting again when the weather turned cold the next fall. Community leaders prepared each meal, and all men took part in plays and dancing. The first spectacle was a play by Marc Lescarbot telling the story of a group of sailors who encounter the god Neptune on their journey to the New World. Guests were also warmly invited to join in the feast and fun. Regular invitees included Sagamore Membertou and members of the Mi'kmaq nation, who were treated as equals. Although idleness is not a cause of scurvy, some of the success of the settlement can be attributed to Champlain's idea — and the healthy eating and genuine comradeship it engendered. Part of the solution were the high-ascorbic strawberries, blackcurrants, and fresh meats that the Mi'kmaq guests brought as presents. Good food, good friends, and an order of good cheer. Dinner theatre was born, and scurvy never stood a chance.

BUTTER TART
The dessert of limited means.

Who says all hospital food is bad? The earliest published recipe for butter tarts appeared in *The Women's Auxiliary of the Royal Victoria Hospital Cookbook* in 1900. Thank you, ladies. Yet this wholly Canadian dessert has its origin many years earlier. The butter tart can be traced back to another group of women – the *filles à marier*, who immigrated to New France as early as 1634. These women – many of them in their teens – were used to making tarts of many kinds back home. Their new home didn't have the same variety of ingredients as the cities and country-side of their old home, so these innovators made do with what they could find, filling their tart shells with butter, sugar, maple syrup, and raisins. Making do with limited means was a way of life in New France – even when it came to dessert.

MCINTOSH APPLE
The well-bred innovation.

From humble beginnings come great things. One day in 1811, while clearing an overgrown plot on his farm in Dundela in what is now eastern Ontario, John McIntosh discovered a fragile sapling of what would become one of the world's favourite apples. The cool nights and clear autumn days of the region had created ideal growing conditions for this hearty, crisp fruit. Farmer John and his family bred and later grafted the tree, producing bushels of its tart and tender apples. The McIntosh entered commercial production in 1870 and soon became the most popular variety in eastern Canada and the northeastern United States. Ideal for both eating and cooking, the McIntosh proved well suited to further grafting, producing at least thirty other varieties in North America and Europe. Associated with qualities of freshness, good health, and simple pleasure, this Ontario innovation was also the nominal inspiration for the Apple Macintosh personal computer, another creation that rose from humble beginnings to dominate its market.

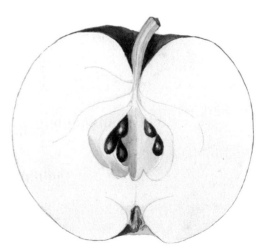

WRINGER WASHER
The first home appliance.

Machines to wash clothes have been around since at least 1791. Yet the washing machine as we know it today can be traced to 1843. In that year, John Turnbull of Saint John, New Brunswick, added a wringer to a washing machine. The top roll of his wringer washer was spring-loaded to rise and fall to accommodate the thickness of the laundry running through the rolls. This mechanism, activated by turning a crank, meant continuous pressure was applied to wet clothes to squeeze as much water out of them as possible. This water would then fall directly into the washer tub. No watery mess. With the device, homemakers could now not only wash clothes but also wring them dry – or at least drier than ever before. Not surprisingly, the wringer washer was a huge hit throughout North America. By 1940, 60 per cent of the 25 million American homes with electricity were equipped with an electric wringer washer – the first home appliance.

RAILWAY AIR CONDITIONING
The traveller's relief.

Climate-controlled cars and buses, planes and trains, homes and offices, stores and restaurants are the norm today. Yet for generations and until quite recently, air conditioning was the exception and not the rule. That doesn't mean the technology hasn't been available. The first air-conditioning system for railway passenger cars dates back to 1859. Credit goes to Henry Ruttan. The enterprising Ontario sheriff used a simple combination of water tank and radiator to generate cool air that made for a much more comfortable ride. He got the idea when he was tackling the problem of railway passenger cars overheating at rest on sidings under the summer sun. Thanks to an Ontario sheriff, while some just cursed the heat, cooler heads prevailed.

CLIP-ON ICE SKATES
The national institution.

Clip-on skates and Confederation: there is an almost perfect symmetry to them. A group of political leaders were responsible for Confederation, which began officially on July 1, 1867. John Forbes of Dartmouth, Nova Scotia, is the man behind the world's first steel ice skates, which earned a patent the very next day. John knew that Mi'kmaq athletes had been playing hockey on the local lakes for centuries, using skates made of bone and leather. As foreman at Starr Manufacturing, John thought he could take that innovation further. He made an all-steel blade that skaters locked onto the soles and heels of their boots by way of a single spring lever. A cultural phenomenon was born: the simplicity and durability of the blade made skating accessible to nearly everyone and helped spur the widespread uptake of hockey in the fledgling nation. The skates were also sold around the world, branded as Acme Skate. (Perhaps Wile E. Coyote owned a pair.) Starr Manufacturing refined the skate's design continually, selling a variety of models for the next seventy years. By then, both skates and the game they inspired were national institutions.

HALF TONE
The printable image.

The modern illustrated magazine was born in Montreal in 1869. Up until that time and place, several systems of photography had been around, but no one had been able to come up with a way to reproduce high-quality photographs as illustrations in magazines. George Desbarats found a way. The latest in a long line of printers, he invented the first half-tone reproduction of photographs. A half tone is the reprographic technique that simulates continuous tone imagery through the use of dots. The dots, which vary in size and in the spaces between them, produce a gradient-like or shading effect that emerges as images. *Half tone* can be used to refer to the process or to images generated by the process. Desbarats put the effect to use beginning in 1869 when he published *Canadian Illustrated News*, the country's first national news magazine. He soon published several other magazines, including *L'Opinion Publique*, *Mechanic's Magazine*, and *Canadian Patent Office Record*. In 1873, he took his creation south of the border and founded the *Daily Graphic*, the first daily newspaper in the United States to feature half-tone illustrations. The printable image was now not only alive, but also international.

HOCKEY
The Canadian game.

You would be forgiven for thinking March 3 is a national holiday in Canada. On that day in 1875, the first organized *indoor* game of hockey was played. It happened in Montreal, to be exact, and involved mostly students of the city's McGill University. Hockey had been played informally several times before that date – both in Canada and in Great Britain. In Canada itself, the historical record contains reports of twenty-four hockey-like games played prior to the big one in 1875. And games in which players wield curved sticks to whack around a ball can be traced all the way back in time to Ancient Egypt. Yet a game contested on ice in a cold building by players wearing sharpened steel blades strapped to their feet is a quintessentially Canadian innovation. Hockey, it seems, is the natural result of frigid winters that cover the country with snow and ice for as few as four and as many as six months every year. Precursors of the modern sport were common among First Nations communities. For instance, Mi'kmaq teams were competing in stick-and-ball matches on frozen lakes near what is now Dartmouth, Nova Scotia, while the first settlers to Halifax were still trying to figure out how to cope with winter in the 1740s. Those early teams wore ice skates made from animal jawbones strapped to their footwear with thongs of hide. Without doubt, hockey has been Canada's game for a long time.

IRONING BOARD
The portable valet.

Not so swish that you can afford to hire a valet to keep your clothes pressed just so? No problem, thanks to John Porter of Yarmouth, Nova Scotia. In 1875, John invented the ironing board. Dead simple — just a board mounted on crossed legs that are hinged at one end so they can be folded out of the way, and the board stowed upright when not in use. Early versions even had a narrow fold-out pressboard for ironing sleeves. The ironing board is a classic example of great innovation: a simple solution obvious to one person where it had never been to others. It's such a clever little item that it has hardly changed since it was first unveiled. Who needs a valet?

BAGGAGE TAG
The traceable luggage.

Even the simplest innovations need to start somewhere. Consider the humble baggage tag. In the early years of rail travel — 1882 to be exact — John Michael Lyons of Moncton, New Brunswick, came up with the idea of baggage handlers writing each passenger's name, departure point, and destination on a separate tag. Each tag would then be torn in two, with the top portion attached to the passenger's bag and the bottom portion kept by the passenger. This simple system made it easy for travellers to find their bags at the end of their trips. It even made it possible for lost baggage to be traced, found, and eventually reunited with its owner. The patent office called this advance in transportation the separable coupon ticket. Today, we all call it the baggage tag, the first traceable luggage.

Harlan's Point.

PANORAMIC CAMERA
The ghosts in the machine.

Have you ever seen an old photograph in which the same person appears twice? You're not seeing a ghost. Well, likely not. The photo was likely taken using a panoramic camera. Developed by John Connon in 1887, the machine is able to photograph a complete circle at one exposure. Its secret was a system of automatic controls that moved the film as the lens took in the gradually changing scene. Revolving atop a tripod, the Elora, Ontario, inventor's new camera could be set to cover any number of degrees without splices. The novelty of the images the camera produced made it popular for many years, especially to capture landscapes and large groups of people such as high school and college classes. Young pranksters quickly learned to stand at one end of their group until the lens had passed by and then sprint to the other end of the gathering to be photographed again — ghosts in the machine.

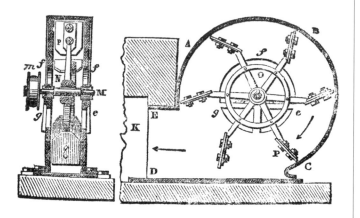

ROTARY VENTILATOR
The cure for the hot tin roof.

We should all age as gracefully as the rotary ventilator. Born in 1889, the ventilator is the creation of James Lipsett of Saint John, New Brunswick. Found in chimneys and ceilings, the ventilator is equipped with blades shaped to use the wind to increase updraft and draw hot air either up a chimney or out of a room or home. The blades are also designed so rain and melting snow don't drip down through them to damage the room, chimney, or the ventilator itself. This simple extractor remains a popular and virtually unchanged element of air exchanger systems found in many homes around the world, alive and turning at more than 125 years of age.

BASKETBALL
The afterlife of a peach basket.

Create a new team sport that demands agility, speed, and accuracy from its players, not just strength alone. Make sure it can be played both safely and indoors. Oh, one more thing: come up with it in fourteen days. James Naismith's answer to his boss's difficult demand was basketball. An instructor at the YMCA training school in Springfield, Massachusetts, in 1891, Naismith drew inspiration for his new game from one he played as a child in the small town of Almonte, just outside Ottawa. That pastime involved young James and a group of his friends throwing stones to knock a small rock off a larger one. To devise a new game, he refined that basic idea by dividing the group of kids into two teams, switching the stones for a soccer-sized ball, and making targets out of two peach baskets nailed high at either end of a gymnasium. Now Naismith had not only a game to satisfy his boss at the Y, but also a sport that would one day be played by millions of people in dozens of countries. That's a lot of peach baskets.

STANLEY CUP
The first team championship.

On the evening of March 18, 1892, members of the Ottawa Hockey Club gathered at the Russell Hotel, situated in the heart of the capital, a short walk from Parliament Hill. During a break in the festivities, a message from Lord Stanley of Preston, Canada's governor general, was read aloud: "I have for some time been thinking that it would be a good thing if there were a challenge cup which should be held from year to year by the champion hockey team in the Dominion. There does not appear to be any such outward sign of a championship at present, and considering the general interest which matches now elicit, and the importance of having the game played fairly and under rules generally recognized, I am willing to give a cup which shall be held from year to year by the winning team." Lord Stanley was true to his word. He purchased a silver bowl for ten guineas. On one side were inscribed the words "Dominion Hockey Challenge Cup." On the other side "From Stanley of Preston." The first team championship trophy in North American sporting history and one of the oldest in the world, Lord Stanley's gift would eventually come to be known simply as the Stanley Cup. Today, it has grown literally and figuratively to become an iconic and beloved symbol of athletic excellence and an instantly recognizable reminder to the world of Canada's high spirit. It's also a significant social innovation — enduring proof of how one man's creative vision and genuine generosity can rally fans, set young hearts to sport, and drive athletes to near-inhuman lengths of courage, stamina, and team spirit. It's the gift that, some 125 years later, keeps on giving.

VISTA DOME CAR
The vroom with a view.

It turned out to be an innovation slightly ahead of its time. In 1896, Canadian Pacific Railway built the first production model of a double-deck mountain observation car. The first vista dome car in the world gave lucky passengers a full, unobstructed view of the incomparable natural splendour of the Canadian Rockies. One problem: the intense summer sun turned the glass-domed cars into ovens. Following the summer of 1897, CPR pulled the cars out of production. The advent of air conditioning in the 1920s made the vista dome car practical, and soon the cars were fixtures on rail lines throughout North America. The time had come.

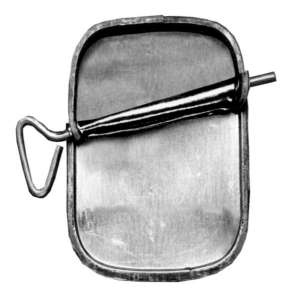

KEY-OPENER CANS
The little door to luncheon meat.

It lived through an entire century in which age-old empires fell, the atom was split, cataclysmic wars were waged, and adventurers journeyed to the moon. Yet it endured. The humble key-opener can was the creation of Joseph Clark of St. George, New Brunswick. Developed in 1900, his can was the first in which one of its sides overlapped and ended in a lip or lug to which a key was attached to open said can. A new century has dawned and key-opener cans abide on grocery shelves around the world. Some things are born perfect.

FIVE-PIN BOWLING
The funner, faster game.

Change the game. That's what Thomas Ryan did when patrons of his ten-pin bowling alley starting complaining. "The ball is too heavy," they said, "and a game takes too long to play. All those pins have to be set up all the time." Rather than try to convince his customers they were wrong (seldom wise), he changed the game. More accurately, he used the old game to create a new one in 1909. The Toronto man's innovation featured a lighter ball, a different scoring system, and fewer and smaller pins — each wrapped in a rubber band to make them easier to topple. Fun and fast, five-pin bowling was an instant success in Mr. Ryan's establishment and quickly caught on across the country. Today, it is played in all provinces and territories across Canada. Canadian Forces Base Alert in Nunavut has the northernmost bowling alley in the world — five-pin, of course.

EGG CARTON
The dimple that settled a fight.

Who says nothing positive ever comes from fighting? In 1911, Joseph Coyle happened upon a heated argument between a deliveryman and a hotelier in his hometown of Smithers, British Columbia. The hotel owner was upset because the eggs shipped from a local farm often arrived cracked or broken. While a newspaper publisher by profession, Joseph was a designer by inclination. The overheard argument inspired him to create the egg carton. The secret of its success is its hard dimples, which protect the carton's delicate contents from the stresses of transport and storage. Hundreds of millions of egg cartons — not much different from Joseph's original creation — have been manufactured and used since. That's a lot of fights that have gone un-fought.

ZIPPER
The hookless fastener.

Many great innovations simply speed up or eliminate the actions that consume our time. The hookless fastener, more commonly known as the zipper, is one of the classics. The man on the other side of the zipper is Swedish-born Gideon Sundback. In 1913, he came up with something he called the Hookless No. 2. It's the metal zipper as we know it today — two strips of teeth brought together tightly by a slider. No more tricky buckles or time-consuming hook-and-eye fasteners. Sundback also created the machine to manufacture his new device and set up the Lightning Fastener Company in St. Catharines, Ontario, to do just that. Surprisingly, use of the zipper didn't strike like lightning. It wasn't until World War Two that Sundback's fastener became popular in the design and making of clothes, and a now-familiar sound truly began to be heard around the world. Zip!

THERMAL WINDOWS
The Canadian must-have.

Few countries have the extremes in temperature Canada has — sweltering heat in summer, painful cold in winter. Generations of Canadians have adapted their homes to manage the back-and-forth swings between the extremes. Thermal windows are one of these adaptations. Created in 1917 by Lawrence McCloskey of Boisetown, New Brunswick, the windows feature two panes of glass separated by a space. According to Mr. McCloskey's design, the edges of the panes are sealed tight and the space filled with alcohol. This interior thermal barrier prevents warm air from escaping and cold air from entering without diminishing or distorting the natural light that passes through the panes. The same design is used to this day. The only difference is that the space between the panes is now filled with argon — a dense, colourless, odourless, inert gas that serves as a happy medium between a nation's extremes.

CANNED LOBSTER
The saviour of the fishery.

Would you eat discoloured lobster meat? Neither would people in 1920. At the time, the unappetizing colour of canned lobster meat threatened to end the multimillion-dollar lobster industry in Atlantic Canada. The country's National Research Council uncovered the source of the problem: bacteria present in the cans produced sulphite of iron, which has a dark colour — harmless yet off-putting. A twin solution was proposed; first, a heat-sterilization process to kill the bacteria, and next a recommendation that canneries add three ounces of germ-fighting vinegar to each gallon of brine used in the canning process. These simple, highly effective interventions put an end to discoloured lobster meat and enabled the industry to claw back its profits for several generations, by which time advanced freezing techniques and refrigerated supply chains lessened the need for canned food.

MULLIGAN
The golfer's little helper.

The next time you slice your drive deep into the woods and then calmly tee up another ball without having to hear a disparaging word from the others in your foursome, you can thank David Mulligan for the honour. While playing at the Country Club of Montreal golf course one day in the 1920s, our hero hit a poor tee shot and without hesitation placed another ball on the tee and took the shot again. Mulligan called his second whack at the dimpled sphere a "correction shot." Golfers the world over would come to call that follow-up shot something else: a Mulligan. Thanks, Dave.

CIRQUE DU SOLEIL
The twenty-first century circus.

In 1984, Guy Laliberté had a vision for the next century's circus. The street performer from Baie-Saint-Paul, Quebec imagined a spectacle that would centre on the virtuosity of the world's finest performing artists. No prancing horses and snarling tigers, s'il vous plaît; no faded big tops or sawdust-strewn floors at local hockey arenas. This new circus would amaze with feats of almost inhuman strength and agility, dazzle with colourful costumes and brilliant lights, captivate with original music played live and continuously, invoke the imagination, provoke the senses and evoke the emotions. This new circus would blaze with the energy of the sun. Fired with the support of key partners, generous professional assistance and timely financial backing from governments, Laliberté made his vision real. Cirque du Soleil grew rapidly from a single touring show to become the largest theatrical producer in the world – staging dozens of immersive spectacles that transfix those in attendance. Artistic triumph has led to business success. Cirque du Soleil now employs four thousand people from fifty different countries performing in tents, theatres, nightclubs, casinos, films and television specials. Over 160 million spectators have experienced Cirque du Soleil – one Canadian's startling vision realized for the world to see.

GROUP OF SEVEN
The Algonquin school.

Seven men gave Canada a powerful, enduring image of itself. Franklin Carmichael, Lawren Harris, A.Y. Jackson, Frank Johnston, Arthur Lismer, J.E.H. MacDonald, and Frederick Varley were their names. Inspired by the paintings of Tom Thomson and his renderings of the northland bush and lakes of Algonquin Park, this group of seven Canadian artists were instrumental figures who through their paintings defined a nation for itself and the world. Before they came along, Canadian artists were Canadian in name only – their work reflected the scenes, influences, and practices of Europe. Thomson and the Group of Seven changed that. They set out individually and collectively to create a wholly Canadian art inspired by and reflecting the lands, forests, and waters of their home and native land. While criticized by some for falsely depicting Canada as largely uninhabited, their common focus on the land and not the population allowed them to experiment with colour and texture in a way that riveted the images of Canada in the minds of people all over the world. When they held their first exhibition in Toronto in 1920, the country's artistic elite greeted their work with scorn. Yet those views were soon cast aside as their fellow Canadians and others throughout the world saw in this work an original and authentic vision of Canada. The names of a smattering of their paintings alone tell the tale of their ambition: *Sumacs*, *Red Maple*, *October Gold*, *The Lumberjack*, and *North of Lake Superior*. That vision continues to resonate strongly among Canadians as the defining expression of the beauty of their country and the character of its people.

WOOD-TILE CROSSWORD
The pre-Scrabble.

What do you get when you combine checkers and a crossword puzzle? The answer is wood-tile crossword. Created in 1926 by Edward McDonald of Shediac, New Brunswick, the new game was played on a checkerboard by two players. Each player had a complete coloured set of lettered game pieces. The object of the game was for players, taking turns, to place their pieces square by square to form words. Sound familiar? You're right. Only twelve years later, Scrabble arrived, with its letter values, bonus squares, letter blanks, and seven-tile racks as further innovations that challenged their playful Shediac predecessor.

WATERPROOF SHOES
The puddle jumper's ally.

The secret is the tongue. Charles Grant designed his waterproof shoe with a rubber-coated tongue that is wedded to the shoe during vulcanization. Vulcanization is a process in which rubber is hardened by treating it with sulphur at a high temperature. You might not think his move is such a big deal these days. That's because Grant's secret is now the standard.

EASY-OFF

The automated elbow grease.

An innovation doesn't necessarily have to upend the world. Sometimes all it needs to do is make one person's life a little bit easier. In 1932, that one person was Herbert McCool. The Regina, Saskatchewan, handyman spent the early years of the Depression fixing stoves and cleaning ovens. Not exactly an effortless occupation, especially the latter task, so he looked for a way to make it easier. Herbert came up with a method to make caustic acid — a compound known for its cleaning powers — stick to the top, bottom, and sides of an oven where it could melt away baked cooking oils and debris. Herbert's wife, Doris, started making and packaging the easier elbow grease, and then he sold it door to door. It worked hard, and its name — Easy-Off — was an inspired choice.

SARDINE TIN

The lunchbox companion.

Fish in a lunchbox on a warm day was never an appetizing option. Until 1932, that is. That year, Henry Austin developed the sardine tin. Why Henry? His brainwave might have had something to do with the fact that the largest sardine plant in the world — Connor Bros. Ltd. — was in his New Brunswick backyard, so to speak. His handy sardine container was small enough to carry in a lunchbox, held just enough of the little devils to make for a satisfying snack, and, conveniently, came with its own detachable key to roll open the tin. The Canadian innovation was soon in use all over the world and made a global trade in sardines possible for the first time. All because Henry wanted fish for lunch.

TABLE HOCKEY
The makeshift holiday gift.

What can you give your kids for Christmas when you don't have any money? That's the question Donald Munro faced at the height of the Depression in December of 1932. Penniless, the Toronto Mr. Fix-It decided to make his family something novel. With wire hangers, clothespins, clock springs, wood dowels, and scrap metal, Munro built a mechanical game of hockey that sat right on the kitchen table. Planning to use a ball bearing as a puck, he built the surface of the ice with two gently sloping halves meeting at an elevated centre. That way, the puck always rolled back into play. Even better, unlike the solo arcade games of the day, Munro's game required one player at either end, pitting one team against another just like in the NHL. The gift proved wildly popular with the Munro kids and their friends, so much so that its amazed creator resolved to go commercial. He convinced Eaton's, the pioneering Canadian department store, to take several games on consignment. Before long, the retailer demanded more—plenty more. Munro turned his idea into gold by starting a company to manufacture a slicker version equipped with components and innovations to more accurately simulate the real game. Creative solution meets pressing need and a business is born. He shoots, he scores.

SUPERMAN
The first hero.

Superheroes are so common these days — on television and at the movies; in comics, books, and apps; alone and in teams — you would be forgiven for thinking they had always been around. Joe Shuster would tell a different tale. In 1933, the Toronto artist and his writing partner, Jerry Siegel, were the first to create a comic book superhero — a mysterious figure who uses extra-ordinary physical powers to uphold good and fight evil. Yet they had to push their creation to publishers for nearly six years — suffering

rejection after rejection — before Superman finally stepped out of the telephone booth in Action Comics No. 1 in June 1938, cover price ten cents. The next year, their Man of Steel earned his own series and took off like a speeding locomotive, selling more than half a million copies per month. The golden age of comic-book superheroes was underway — a welcome tonic to a world on the brink of the most destructive conflict in human history. Today, these heroes and heroines are just as popular. In 2014, a single copy of Action Comics No. 1 sold at auction for $3.2 million — an approving nod to Canadian Joe Shuster's first hero.

NATIONAL FILM BOARD
The Canadian storyteller.

When you sit just north of the world's most powerful purveyor of filmmaking creativity, technological genius, and sheer movie-producing volume, you have to take a dramatic step to ensure the survival of your own filmmaking identity. Founded in 1939, Canada's National Film Board (NFB) was that step. The NFB was the world's first organization dedicated solely to producing and distributing documentary films, animated films, and alternative dramas that reflect a specific country's experiences and aspirations. Over its history, the organization has been responsible for more than 13,000 productions and has earned some 5,000 honours, including 12 Academy Awards. Revered by film lovers around the world, this national institution dedicated to film also served as the inspiration and guide for Ghana when that country set up the Gold Coast Film Unit. One country's dramatic innovation sparked another's.

SYNTHESIZER
The sound of oscillation.

If you're a fan of Bjork, Avicii, Radiohead, or any other artist whose music relies on electronic sounds, say a word of thanks to Hugh Le Caine. In 1937, the Canadian physicist created the world's first music synthesizer. He called his creation the electronic *sackbut*, a nod to an ancient form of trombone. Hugh's sackbut is a device that applies technologies used in radar, radio, and atomic physics — filters, waveform generators, frequency and amplitude modulators — to generate and modulate sound. Virtually anyone can play his instrument (playability was one of Hugh's two prime considerations when developing an instrument, the other being that it must make beautiful sounds). The right hand presses the keys on the keyboard while the left hand manipulates a wheel to control the volume, pitch, and timbre of each note played. Today's synthesizers, which feature several control dials and switches, can be traced to Hugh's creation.

CANADIAN BACON
The protein in uniform.

If you love eating back bacon, thank the British. Well, half thank them. Great Britain was cut off from its regular European suppliers of food during World War Two, so the Brits turned to their allies in Canada for help. Along with sending many tonnes of eggs, egg powder, and poultry, Canadians developed and shipped a special kind of bacon. You see, British taste for bacon differed from the *streaky* strip bacon enjoyed by North Americans. Our cousins across the Atlantic preferred slices. To satisfy the British yen, Canada's National Research Council developed an oblong-shaped slab of bacon that could be cut into slices. Some two tonnes of what came to be known as Canadian bacon were shipped across the Atlantic each week for the duration of the war: protein in uniform, sailing into battle.

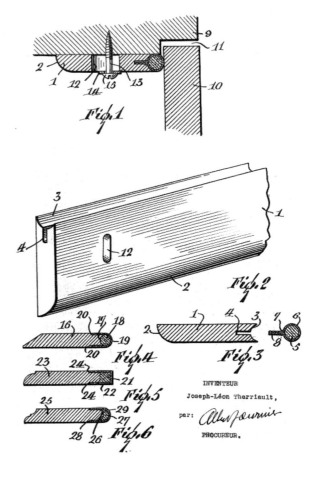

WEATHERSTRIP
The end of the draft.

Scarcity often inspires innovation, and scarcity was a way of life during World War Two. In 1943, wartime restrictions on metal and rubber inspired Joseph Therriault to create a new kind of weatherstrip. The New Brunswicker's insulator combined a wood moulding with a cushion strip made out of felt. He also included a series of slots for screws. Homeowners and handymen could use the screws to tighten or slacken Joseph's strips against their window frames. So while Canadians were consumed with questions surrounding a wartime draft, Joseph quietly ended a different draft closer to home.

ENLIGHTENMENT SHORT STORY
The revealing moment.

With the publication of her first short story, *The Dimensions of a Shadow*, in 1950, Alice Ann Munro began her professional literary career. Over the next sixty-five years and counting, the Canadian writer revolutionized the ambition and architecture of the short story. Her tales, now numbering in the dozens, trace many subjects: love and work, small towns and big dreams, coming of age and growing old, being with family and being alone, authenticity and ambiguity. Little happens on the surface but much is laid bare. With an attentive eye, as well as an open mind and receptive heart, the reader is exposed to a ray of enlightenment – word by word, page by page, drop by drop, as the stories move back and forth in time to reveal their truths. Alice Munro's new approach was adopted by writers around the world, and especially in Hollywood where filmmakers and their audiences took a page out of the Canadian writer's books and abandoned their insistence on strict chronology. Critics shared the enthusiasm, as three Governor General's Awards, a Giller Prize, a Man Booker International Prize, and a Nobel Prize in Literature make clear.

NANAIMO BARS
The nation's favourite.

Legend has it the 1953 edition of Edith Adams's cookbook featured a new treat in the dessert section. This one was easy to make, yet looked complex: a three-layer delight with a packed-crumb base, hard chocolate top, and creamy custard middle. To fabricate an endearing heritage for the distinctive bars, they were named after a much-admired small town on Vancouver Island – Nanaimo. Little did Edith know she had created a sweet sensation. The bars soon became a staple of busy Canadian cooks, who served the dessert not only at the family dinner table but also at luncheon get-togethers, church suppers, and community gatherings everywhere. Today, Nanaimo Bars are a Canadian institution, repeatedly voted the country's favourite confection. Gourmands around the world are also discovering this sweet treat. It can be found in British shops, Manhattan cafés, and coffeehouses in Laos, Spain, Taiwan, and even Australia – about as far from Nanaimo as a sweet tooth can get.

POUTINE
The delicious damned mess.

What happens when you combine French fries, cheese curds, and steaming hot gravy? According to Fernand Lachance, you get *une maudite poutine*, a damned mess! Many believe the proprietor of Café Ideal in Warwick, Quebec, served up his first dish of poutine in 1957 when a patron asked him to scatter cheese curds over his customary order of fries. Fernand did as asked, adding two practical innovations: he served the jumble on a plate to keep it in one place and covered the whole thing with gravy to keep the dish warm. By the 1980s, his creation had spread throughout Canada and into the northeastern United States. It became a staple in local restaurants and on the menus of the world's biggest fast-food chains. Franchises devoted to the dish sprang up to add all manner of toppings and twists to the traditional recipe. Now poutine is poised to conquer taste buds farther afield. In 2016, poutine was served at a White House state dinner in honour of Canada. In Russia, they routinely serve a local version called Raspoutine. *Félicitations*, Fernand, your poutine is a damned mess the world over.

CANADIAN DICTIONARY OF BIOGRAPHY
The history of a people.

While not singular in concept, the *Canadian Dictionary of Biography* is unique in structure. Unlike the British and American versions, which list biographic entries alphabetically, the Canadian edition publishes by period, with each volume covering a specific range of years. Biographies within each volume are arranged alphabetically. This time-period approach has three main advantages: it brings together scholars who specialize in certain eras, making researching, editing, and cross-checking entries easier; it removes the need for readers to reacquaint themselves with the historical period in which the depicted individuals lived; and it allows revisions to be made to published volumes without having to update the entire series. Begun in 1959, the *Canadian Dictionary of Biography* is also a bilingual publication of the University of Toronto and Université Laval. The first volume of this distinctive dictionary appeared in 1966, featuring 594 biographies from the years 1000 to 1700. Today, its fifteen volumes include 8,419 biographies that take the entire publication up to 1930 — with more on the way soon. Impressive in scope and entirely unique in structure, the *Canadian Dictionary of Biography* is the history of many persons that tells the incredible story of one people.

GOULD'S MUSICAL TECH
The studio genius.

Glenn Gould's performance career ended in Los Angeles on April 10, 1964, the day of his last live concert. The Canadian pianist's artistic life didn't end then and there, though. On the contrary, it moved to a different and, some would argue (perhaps Gould himself), higher plane. In the recording studio and in his writing and broadcasts related to that work, Gould probed the deepest questions known to artists: What is art? What defines creativity? Is a recording any less authentic than a live performance? His recording experiments and philosophy challenged conventional responses to these questions. He forced us to look much more closely at our understanding of the relationship among composer, performer, and listener. He inspired us to reconsider our predispositions about originality and authenticity. In the move from stage to studio, this genius paradoxically redefined the very nature of artistry itself.

WONDERBRA
The power of the push-up.

Some products are so potent their names have defined a category. Think Kleenex for tissue, Teflon for non-stick, and Vaseline for petroleum jelly. The same goes for Wonderbra and women's intimate apparel. Wonderbra was the 1964 brainchild of designer Louise Poirier of the Canadian Corset Company. The product's name hinted at the revolution her brassiere would spark in an age when four women in ten still wore girdles. Until then, women's undergarments were mainly new spins on old-fashioned corsets and girdles. To its credit, the Montreal business realized modern women wanted something more in line with the era's revealing new fashions as well as the new-found independence of women themselves – something feminine yet freeing, stylish yet supportive. Wonderbra fit the bill. Featuring fifty-four design elements, the undergarment is a lacy brassiere based on a push-up construction that "lifts and supports to comfortably create a fuller form" – qualities that marketing research suggested modern women wanted. It was indeed a hit. By 1979, Wonderbra dominated the Canadian market. Women lined up to buy it when it was later released in the United States. Wonderbra remains a favourite around the world, a wearable symbol of social freedom that brought the women's undergarment business out of the closet, launched the new multi-billion-dollar intimate-apparel industry, and earned the brand loyalty of countless women.

FLAG COLOUR STANDARDS
The 500,000 shades of red.

A red maple leaf on a white background between two red bars is recognized around the world as the Canadian flag. Even more, its vivid, contrasting colours symbolize the bold spirit of the people that make up the country. Now imagine that same maple leaf and bars in washed-out orange and that stark white background as dingy grey. Not a pretty picture. Yet that's what many of the first flags flown in 1965 looked like. The true north may be strong but its flag was suffering in the nation's harsh climate. Enter the researchers at Canada's National Research Council. They tested fabric samples and dyes for their strength, exposing them to the elements and to the organization's wind tunnels. They then used the results of these rigorous trials to create the first standard for the national flag of Canada. It identified everything from the types of fabric, grommets, and sewing thread to be used to the ideal colour red – selecting the best from among half a million shades. It marked the first occasion in which international standards for colour were applied to a national flag. Equipped with this clear guide, manufacturers across the country could now produce flags as strong and bold as the country itself. The maple leaf forever.

SPHYNX CAT
The hairless cuddle.

When Riyadh Bawa heard rumours of a kitten being born without hair, he didn't see a mere novelty. He saw an opportunity. The year was 1966 and the University of Toronto science student believed he could take advantage of the recessive gene within that young cat to develop a new breed for people who were allergic to furry felines. So he bought the kitten and its mother and – with the help of his own mother (a breeder of Siamese cats) – mated the two cats to produce a litter of hairless kittens. He then bred the males of the litter with American shorthair females to produce what became known as the Sphynx. The name is a bit of a misnomer as there isn't much mystery about these cats. They love to get up close and personal due to the fact they lose body heat much more quickly than their hirsute cousins. Call them the hairless cuddle.

YUKON GOLD
The world's spud.

European and South American immigrants to southern Ontario couldn't understand why there were no yellow-fleshed potatoes in their adopted country. The sunny spuds were a staple back in their home countries, yet all they could find in the neighbourhood markets of 1950s Canada was a pale-fleshed, flavourless imitation. University of Guelph research scientist Gary Johnston wasn't convinced of the need for a yellow-fleshed potato until he sampled one for himself – straight from the Peruvian plantation of a graduate student's father. One taste changed the good doctor's mind, launching him on a two-decade quest to create the perfect potato. Crossing several domestic and foreign varieties, in 1966 he finally evolved a new breed that featured smooth, eye-free skin and tasty flesh. It could be boiled, roasted, baked, or fried and, best of all, readily grown in the cooler climes of North America. The name he and colleague Norman Thompson devised paid tribute to the spud's distinctive hue and Canadian patrimony – Yukon Gold. It has since become a culinary superstar, grown in Japan, adored in Finland and Sweden, and found on menus throughout the world.

GREAT SOUND
The world's first audio standards.

What is great sound? Until 1970, the absence of any accepted way to measure the performance of loudspeakers led to inconsistencies in quality and design, unfounded claims by manufacturers, and consumers left adrift. That year, Canada's National Research Council set out to define great sound. A research team built an anechoic chamber—a room that absorbs sounds and eliminates echoes—then placed a speaker in the chamber and tested the sound it produced at different spots in the room. The team did the same with speaker after speaker to measure the performance of each. Floyd Toole, one of the organization's researchers, developed a further series of precise tests. One involved asking listeners sitting in a mock living room to rate identical sounds reproduced by various speakers hidden behind a thin curtain. Listeners rated the sounds according to clarity, fidelity, definition, fullness, brightness, and overall pleasantness. The Audio Engineering Society picked up on Floyd's work, publishing it and attracting attention from manufacturers, audiophiles, and industry magazines. Floyd's audio standards led to a revolution in speaker design, as designers understood with crystal clarity the performance characteristics of great sound.

BOVRIL
The broth with beef.

"Send me one million cans of beef." Tall order. Yet that was exactly the command made by Napoleon III in 1870. The French emperor needed to feed his troops during their battle against Prussian forces. John Lawson Johnston, a Scotsman living in Canada, was up to the task. But instead of canning meat, he created a fluid beef extract he called Bovril from *bov* meaning *ox* and *vril* meaning *strength*. The product capitalized on the large quantity of beef available in Canada. Johnston's sterilized extraction and canning process made it possible to store the cans for extended periods of time and transport them great distances without spoiling. Since then, the so-called King of Beef (Bovril, not Napoleon) has become popular around the world — in war and peace, among French soldiers, Arctic explorers, and English soccer fans, for cooks and those who just want a quick, warm, nourishing pick-me-up. Strong as an ox? I'll take a million.

DERIVATIVE VALUATION
The optional formula.

A derivative is a contract whose value stems from the performance of something else, usually an asset that an investor will buy or sell at a later date. How to value these assets has plagued generations of economists. The tricky element in any valuation is risk. Assigning a risk premium is difficult because assessing an investor's attitude toward a risk, while easy in theory, is a much tougher nut to crack in practice. The Black–Scholes formula cracked this nut. Developed in 1973 by Fischer Black of Washington, D.C., and Myron Scholes, who was born in Timmins, Ontario, the formula is a mathematical model of a financial market that contains derivative-investment instruments. The implications of the model are far from theoretical. Overnight, the American-Canadian team's work legitimized the activities of options markets and propelled the volume of options trading to new heights. The Chicago Board Options Exchange (CBOE) was able to open the same year as the formula was made public and within a single decade was trading over one million options contracts each year. So revolutionary is their work that Dr. Black — the American — and Dr. Scholes — the Canadian — were awarded the Nobel Prize in economics. Nothing complicated about that.

MULTI-TOUCH SCREENS
The pinch and the zoom.

Innovators had the idea for multi-touch screens in their minds and down on paper for years. The true breakthrough in this technology came in 1982. It occurred at the University of Toronto when members of the school's Input Research Group actually made the first human-input multi-touch screen. Their screen featured a frosted-glass panel with a camera behind. The camera detected when a finger or fingers were placed on the panel and registered these input points as black spots on a white background. This practical application of plural-point awareness released a cascade of further research and design. It was first most evident in the movies and TV shows we watched. Now it's a fixture on our mobile digital devices, as we pinch and zoom into the future.

JUSTE POUR RIRE
The world's comedy festival.

Montreal is the world's capital of funny. Each summer, the city attracts dozens of top stand-up comics, more than a thousand of the most powerful entertainment executives, and some two million comedy fans from around the globe. The gathering place for them all is Juste Pour Rire. It started in 1983 as a modest two-day event. Two years later it debuted in English as Just For Laughs. Since then, founders Andy Nulman and Gilbert Rozon have taken the world's first annual comedy festival and grown it into a multimedia empire. *Juste Pour Rire* and *Just For Laughs* TV shows are broadcast in more than 140 countries and are available on some 100 airlines around the world. The nearly 300 episodes of *JPR Gags*, which account for more than 1.5 billion hits on YouTube, are the largest collection of non-verbal hidden-camera comedy ever filmed. On top of all that, the festival itself grows in size and popularity each year. The capital of funny keeps getting funnier.

INTERNET SEARCH ENGINE
The archive without a v.

Before Yahoo! and Google there was Archie. The world's first search engine, Archie was a project of students at Montreal's McGill University School of Computer Science, built to connect them to the Internet. Alan Emtage wrote the earliest version of the search engine in 1990. His program compiled a list of File Transfer Protocol archives that were stored in local files. Bill Heelan and Peter Deutsch then wrote a script that enabled users to log in and search those files. Within two years, Archie contained some 2.6 million files that amounted to 150 gigabytes of data. While an impressive feat at the time, Archie – shortened by one letter from the word *archive* to fit within character limits – is now a museum piece. It rests in Poland, a legacy of the earliest years of the Internet.

3D SCANNING
The image in the matrix.

Virtual reality has long been a dream of engineers. For years, however, attempts to realize the dream were clearly seen as synthetic computer generations. In 2004, engineers at Canada's National Research Council made the dream real — or as close to real as virtual can get. In that year, the organization brought together the latest advances in data storage, accelerated processing, and desktop applications to create 3D scanning. The advances in storage and processing were especially crucial, as the scanning process often involves millions of measurements that produce equally large volumes of data. Unlike the cartoon-like reproductions of primitive virtual reality, 3D scanning generates precise and realistic visualizations that reproduce real life in almost microscopic detail. [...]rs of the *Lord of the Rings* and *Matrix* movie franchises were quick to jump, using the technology to replicate scenes and stunts that are humanly impossible. Closer to ground, French art restoration and preservation professionals took advantage of the advance to create the first archival-quality 3D virtual model of Leonardo da Vinci's *Mona Lisa*, smile and all.

11.5

11.0

10.5

10.0

9.5

9.0

8.5

8.0

7.5

7.0

6.5

6.0

5.5

5.0

4.5

4.0

3.5

3.0

2.5

2.0

1.5

1.0

0.5

Afterword

Each of us has the power and the duty to imagine better ways, better things, better times.

Canadians have a knack for innovation, and our prosperity as a nation is more dependent than ever on fresh ideas. It is the authors' fondest hope that after reading this little book, some Canadians who have not yet thought of themselves as ingenious might have the courage to do so.

Join us in our quest for new ways to think and act that will continue to make the world smarter, smaller, kinder, safer, healthier, wealthier, and happier. Could there be any better way to spend our time together?

A Timeline of Innovation

EARLY HISTORY

Buffalo Coat

Canoe

Dogsled

Duck Decoy

Igloo

Kayak

Lacrosse

Life Jacket

Longhouse

Maple Syrup

Megaphone

Moccasin

Potlatch

Snow Goggles

Snowshoes

Toboggan

1600s

1606   Dinner Theatre

1634   Butter Tart

1700s

1784   Ships' Knees

1800s

1801   Red River Cart

1811   McIntosh Apple

1833   Ship's Propeller

1837   Fish Ladder

1839   Diver's Air Tank

1840   Canadian Aboriginal Syllabics

1843   Wringer Washer

1844   Pulp and Paper

1853   Compound Steam Engine
Foghorn

1854   Kerosene
Odometer

1857   Oil Drilling

1859   Railway Air Conditioning

1861   Greenback

1862   Oil Pipeline

1863   Jerker Line

1867   Canada
Clip-On Ice Skates

1868   Steam Buggy

1869   Half Tone

1870   Bovril
Rotary Snowplough

1872   Lubricating Cup

1874   Light Bulb
North-West Mounted Police
Telephone

1875   Hockey
Ironing Board

1878   Handset

1879 Roller Bearing

1880 Hot and Cold Faucet

1882 Baggage Tag
Electric Range

1883 Trolley Pole

1884 Peanut Butter
Standard Time

1885 Chicken Bone
Screw Link
Transcontinental Railroad

1886 Experimental Farm

1887 Panoramic Camera

1889 Rotary Ventilator
Syndicated Journalism

1891 Basketball
Marquis Wheat

1893 Stanley Cup

1894 Brunton Compass
Caulking Gun

1896 Interest Calculator
Vista Dome Car

1899 Developing Tank

1900s

1900 Brownie
Disappearing Propeller
Key-Opener Cans
Radio Voice Transmission

1901 Atomic Recoil
Caisse Populaire

1903 Commercial

1904 Canada Dry

1905 National Atlas

1906 Movie Theatre

1907 Russel Logging Boat

1908 Nuclear Physics
Robertson Screw

1909 Five-Pin Bowling
Retail Cosmetics
Silver Dart
Superstardom

1910 Jolly Jumper

1911 Egg Carton
Quick-Release Buckle

1912 Crispy Crunch

1913 Zipper

1914 Rotary Car Dumper

1915 Gas Mask

1916 National Research Council

1917 ASDIC
Curtiss Canuck
Straw-Gas Car
Thermal Windows

1919 Back-up Light
Buckley's Mixture

1920 Canned Lobster
Chocolate Bar
Dump Truck
Forensic Pathology
Group of Seven
Liquid Helium
Mulligan
Waterproof Shoes

1921 Insulin

1922 Documentary Film
Oil Can

1925 Electric Radio
End of Grain Rust
Mount Logan Barometer
Snowblower

1926 Law of Absolute Zero
Wood-Tile Crossword

1927 Monitor Top Fridge

1928 Variable-Pitch Propeller

1929 Rod Weeder

1930 Rib Sheers
Road Lines
Streamlined Locomotive
Whoopee Cushion

1931 Montreal Procedure
Pablum
Plexiglas

1932 Easy-Off
Sardine Tin
Table Hockey
Tempered-Steel Rails

1933 Superman

1935 Snowplane

1936 Atlas of the Heart

1937 Blood Transfusion Service
Self-Propelled Combine Harvester
Snowmobile
Synthesizer
Walkie-Talkie

1938 Electron Microscope

1939 Coffee Crisp
National Film Board
Paint Roller
Shreddies
Stars in Globular Clusters

1940 Aircraft Mass Production
Canadian Bacon
G-Suit
Project Habakkuk

1941 Hormone Treatment

1942 Surgical Curare
Weasel

1943 Weatherstrip

1946 Air Ambulance
Uneven Incentives

1947 Accessible Bus
Smokejumper

1948 Declaration of Human Rights
Snow

1949 De Havilland Beaver
Orenda

1950 Avalanche Protection
Enlightenment Short Story
Garbage Bag
Hodgkin's Cure
Neutron Scattering
Roto Thresh Combine Harvester

1951 Cancer Bomb
Electron Transfer Theory
Pacemaker

1952 Alouette Satellite
Chemical Bridge
Flexi-Coil Air Seeder
Stem Antenna

1953 Electric Wheelchair
Nanaimo Bars

1954 Spiral Nail

1955 Alkaline Battery
Instant Replay

1956 Beartrap

1957 Multiplex Cinema
Peacekeeping
Poutine
Scarborough Suitcase

1958 Avro Arrow
Reaction Dynamics
Shrouded Tuyere

1959 Black Brant
Canadian Dictionary of Biography
Crash Position Indicator
Goalie Mask
Molecular Spectroscopy
Reverse Osmosis
Ski-Doo

1960 Instant Mashed Potatoes
Media Studies

1961 Euro
Stem Cells

1962 Microsurgical Staple Gun
Plate Tectonics

1963 Aids for the Blind

1964 Gould's Musical Tech
Wonderbra

1965 WATFOR

1966 Flag Colour Standards
Sphynx Cat
Yukon Gold

1967 Multi-dynamic Image
Particle Physics

1969 Bloody Caesar
Confederation of Canadian Unions
Digital Photography

1970 Air Seeder
Great Sound
Key Frame Animation
Sorghum Decorticator
Thermofloat Coat

1971 CANDU Reactor
IMAX
Prosthetic Hand

1972 Computerized Braille

1973 Dendritic Cell
Derivative Valuation

1974 Canola
Laser Dinghy
Restorative Justice

1975 Digital Telephone Switch
Second Cup
Saturday Night Live
Telomeres

1977 DNA-Based Chemistry

1978 Visual Neurophysiology
Yuk Yuk's

1980 Tutti-Frutti Modelling Dough

1981 Canadarm

1982 Catalytic RNA
Meningitis Vaccine
Multi-touch Screens
Trivial Pursuit

1983 Blue Box Recycling
HIV Cocktail
Juste Pour Rire

1984 Bomb Sniffer
T-Cell Receptor
Cirque du Soleil

1985 Sulcabrush

1987 Loonie
Pealess Whistle
WEEVAC

1989 ACTAR 911
Language Theory
Oxford Online

1990 Holistic Aircraft Inspection
Ice Cider
Internet Search Engine
Solid Honey
Xylanase

1992 Breakfast for Learning
Space Vision

1993 56K Modem
Rapid HIV Test

1995 Java

1996 BlackBerry
Cradleboard
Two-Way Messaging

1997 Argan Oil Cooperative

1999 Me to We
Nunavut
Zombie Stars

2000s

2000 Growing Nail
Right To Play

2001 Neutrino Mass

2004 Colour Coins
3D Scanning

2005 Climate Rights
Ecotraction
Hydrokinetic Turbine
Miovision

2008 Abeego
Massive Open Online Course
Truth with Reconciliation

2009 Intelligent Teapot

2010 HerSwab

2011 Cattle Plague Vaccine
Helicopter Cushion

2012 Wound Diagnostics
Xagenic X1

2013 iTClamp
Telesurgery
Art as Innovation

2014 Milk Carton 2.0

2015 Homeless Hub
SakKijânginnatuk Nunalik
Very Early Language Learning

2016 Inuit Arctic Research
No-Repair Bridges

ACKNOWLEDGEMENTS

The authors wish to express their delight and appreciation for the expert advice of the many collaborators across Canada who made *Ingenious* possible. They include . . .

Maria Aubrey, Christine Balasch, Alex Benay, Guy Berthiaume, Derek Beselt, Cynthia Biasolo, Dick Bourgeois-Doyle, Catherine Campbell, Maria Cantalini-Williams, Joanne Charette, Elizabeth Chestney, Azka Choudhary, Lois Claxton, Annabelle Cloutier, Sandra Corbeil, Stephen Downes, Joe Dwyer, Jacob Dwyer, Matthew Dwyer, Tessa Dwyer, Carol Elder, Guy Freedman, Chad Gaffield, Daniel Goldberg, Jean Paul Gladu, Scott Haldane, Brian Hanington, Brent Herbert-Copley, Kimberlee Hesas, Ted Hewitt, Monique Horth, Caroline Jamet, Millie Knapp, Jean Lebel, Steven Leclair, Joe Lee, Hélène Létourneau, Laurie Maier, Soriana Mantini, Richard Mayne, Ryan McKay-Fleming, Craig McNaughton, Marcia Mordfield, André Morriseau, Duncan Mousseau, Sheila Noble, Andrew Norgaard, Gilles Patry, Doug Pepper, Luiza Pereira, Leanne Perreault, John Phillips, Sarah Prevette, Neil Randall, Tony Reinhart, Julie Rocheleau, Fiona Smith-Hale, Wilf Stefan, Renée Tremblay, Lahring Tribe, Margot Vanderlaan, Paul Wagner, Stephen Wallace, Christopher Walters, Tonia Williams . . .

. . . and all the other creative souls whose thoughtful work brought this volume to life.

IMAGE CREDITS

particle physics (29) Carlos Clarivan/Science Photo Library

visual neurophysiology (30) ARZTSAMUI/Shutterstock.com

hydrokinetic turbine (31) photograph courtesy of New Energy Corporation, Inc.

neutrino mass (32) photograph courtesy of Lawrence Berkeley National Laboratory

zombie stars (33) photograph courtesy of NSERC

WATFOR (34) image courtesy of The University of Waterloo

Xagenic (34) image © Xagenic Inc.

very early language learning (35) photo courtesy of University of British Columbia Infant Studies Centre/ Martin Dee

Inuit arctic research (36) photograph © Rodd Laing

Massive Open Online Course (37) photograph © Axel Pettersson

odometer (45) Mr Doomits/Shutterstock.com

Canadian aboriginal syllabics (46) Image courtesy of Victoria University Library (Toronto)

Euro (62) Westend61 Premium/Shutterstock.com

Canadarm (64) Handout/Getty Images

language theory (65) Richard Lautens/Getty Images

56K modem (66) Artur Debat/Getty Images

Java (66) McIek/Shutterstock.com

Nunavut (68, 84) Wayne Lynch/All Canada Photos

potlatch (70) photograph courtesy of UBC Museum of Anthropology/Kyla Bailey

longhouse (71) photograph courtesy of Sainte-Marie Among the Hurons, Midland, Ontario

fish ladder (71) David Gowans/Alamy

long johns (72) Hulton Archive/Getty Images

straw-gas car (72) photograph courtesy of University of Saskatchewan Library, University Archives & Special Collections, Photograph Collection A-2926

forensic pathology (73) photograph courtesy of RCMP Heritage Centre, Historical Collections Unit

paint roller (74) Sashkin/Shutterstock.com

accessible bus (74) photograph courtesy of Walter Callow Wheelchair Bus

Confederation of Canadian Unions (78) image courtesy of Confederation of Canadian Unions/Sean Cain/A. G. Nakash/Lea Roback Foundation

Restorative Justice (79) photograph of Mark Yantzi courtesy of Conrad Grebel University College

Restorative Justice (feather) (79) Potapov Alexander/ Shutterstock.com

blue box recycling (80) Andrew Park/Shutterstock.com

solid honey (81) irin-k/Shutterstock.com

Argan Oil Cooperative (82) Education Images/Getty Images

Breakfast for Learning (82) Valentina Razumova/ Shutterstock.com

Cradleboard (83) image courtesy if Saskatchewan

Cradleboard Foundation; University of Saskatchewan and Nehewin Foundation

Abeego (85) image © Kelly Brown

Me To We (85) Colin McConnell/Getty Images

Milk Carton 2.0 (86) image © Missing Children Society of Canada

Right To Play (87) photograph © Right To Play

Truth With Reconciliation (88) The Canadian Press/Adrian Wyld

life jacket (92) Pitchayarat Chootai/Shutterstock.com

diver's air tank (94) Niv Koren/Shutterstock.com

kerosene (95) fatgallery/Shutterstock.com

North West Mounted Police (97) image © Canadian War Museum

hot and cold faucet (98) RedlineVector/Shutterstock.com

trolley pole (98) Tumarkin Igor/ITPS/Shutterstock.com

quick-release buckle (100) JackK/Shutterstock.com

back-up light (102) Sunny_Images/Shutterstock.com

road lines (103) SARIN KUNTHONG/Shutterstock.com

smokejumper (105) photograph courtesy of Saskatchewan Ministry of the Environment

avalanche protection (106) photograph courtesy of Revelstoke Museum and Archives

shrouded tuyere (109) photograph courtesy of Air Liquide Canada

Thermofloat coat (111) photograph courtesy of University of Victoria Archives

pealess whistle (112) Sashkin/Shutterstock.com

ACTAR 911 (112) Ververidis Vasilis/Shutterstock.com

Miovision (115) © Miovision Technologies

HerSwab (115) photograph © MaRS Discovery District

peanut butter (120) Science Photo Library/Shutterstock.com

Buckley's (120) offstocker/Shutterstock.com

end of grain rust (121) photograph courtesy of Agriculture and Agri-Food Canada / Government of Canada

Montreal Procedure (123) photograph reproduced by permission of the Osler Library of the History of Medicine, McGill University, Wilder Penfield Fonds P142

atlas of the heart (124) photograph reproduced by permission of the Osler Library of the History of Medicine, McGill University, Maude Abbott Collection P111

Hormone Treatment (126) photograph © University of Chicago Photograph Archives/Archie Lieberman

surgical curare (126) photograph reproduced by permission of the Osler Library of the History of Medicine, McGill University

Hodgkins Cure (127) photography courtesy of the University of Toronto Archives

chemical bridge (130) photograph courtesy of Stanford University

stem cells (131) Claudio Divizia/Shutterstock.com

sorghum decorticator (132) JIANG HONGYAN/Shutterstock.com

CANDU reactor (133) Frank Lennon/Toronto Star/Getty Images

prosthetic hand (134) belushi/Shutterstock.com

dendritic cell (134) image © National Institutes of Health

telomeres (135) Designua/Shutterstock.com

DNA-based chemistry (136) AP Photo/Tobbe Gustavsson

catalytic RNA (136) AP Photo/Bob Child

HIV cocktail (137) DIOMEDIA/Medical Images RM/Andreas Schindler

T-cell receptor (138) The Canadian Press/Nathan Denetter

Sulcabrush (138) image courtesy of Sulcabrush Inc.

rapid HIV test (139) photograph reprinted with permission from The Chronicle Herald

climate rights (140) The Canadian Press/Chris Windeyer

ITClamp (142) photograph courtesy of iTrauma Care

Chickenbone (147) image courtesy of Ganong Bros. Ltd.

interest calculator (149) photograph courtesy of the Western Development Museum, Saskatoon

brownie camera (148) photograph courtesy of Margot Vanderlaan

retail cosmetics (152) gresei/Shutterstock.com

Jolly Jumper (154) photograph courtesy of Jolly Jumper

chocolate bar (155) Drozzhina Elena/Shutterstock.com

rod weeder (156) courtesy of Morris Industries

whoopee cushion (157) Andrew Paterson/Alamy

self-propelled combine harvester (158-159) Cristi Kerekes/Shutterstock.com

Coffee Crisp (160) Felix Choo/Alamy

Shreddies (160) Paul_Brighton/Shutterstock.com

instant replay (161) © Canadian Broadcasting Corporation

multiplex cinema (162) © City of Ottawa Archives

alkaline battery (163) © The Globe and Mail/AP Images

instant mashed potatoes (163) Sergey Lapin/Shutterstock.com

multi-dynamic image (164) Doug Griffin/Toronto Star/Getty Images

bloody Caesar (164) Jeff Wasserman/Shutterstock.com

digital photography (165) Bedrin/Shutterstock.com

air seeder (165) photograph courtesy of the Western Development Museum, Saskatoon, Bechard Collection

key frame animation (166) image courtesy of the National Film Board of Canada

laser dinghy (167) photograph courtesy of Joe Dwyer

Saturday Night Live (168) Allan Tannenbaum/Getty Images

Yuk Yuk's (169) logo © Yuk Yuk's Inc.

trivial pursuit (170) Rawdon Wyatt/Alamy

Loonie (170) © 2017 Royal Canadian Mint. All rights reserved.

Colour coins (170) © 2017 Royal Canadian Mint. All rights reserved. The Poppy, when used as a symbol of Remembrance in Canada, is a registered trademark of Dominion Command of The Royal Canadian Legion and is used with the kind permission of Dominion Command.

ice cider (171) © Robert Galbraith

chocolate bar (155) Drozzhina Elena/Shutterstock.com

moccasin (176) © Ezume Images/Shutterstock.com

butter tarts (177) photograph courtesy of Joe Dwyer

clip-on ice skates (179)

hockey (171, 180-181) Hulton Archive/Getty Images

baggage tag (182) BrAt82/Shutterstock.com

panoramic camera (camera) (183) photograph from the Archives of Ontario, C 286-4-0-4

rotary ventilator (184) Morphart Creation/Shutterstock.com

basketball (184) Bettmann/Contributor/Getty Images

Stanley Cup (185) used with permission of the Hockey Hall of Fame Museum and Archives

key-opener can (186) Ken Tannenbaum/Shutterstock.com

five-pin bowling (186) Digipear/Shutterstock.com

egg carton (187) safakcakir/Shutterstock.com

zipper (187) urfin/Shutterstock.com

Mulligan (189) Everett Collection/Shutterstock.com

Cirque du Soleil (190) photograph courtesy of Cirque du Soleil

Group of Seven (191) photograph from the Archives of Ontario, F 1066

Waterproof shoes (192) PolakPhoto/Shutterstock.com

Wood-tile crossword (192) photograph courtesy of Alvin Richards

hockey player (194) courtesy of the publisher

Superman (195) Hulton Archive/Handout/Getty Images

Canadian bacon (194) whitemaple/Shutterstock.com

weatherstrip (197) © Innovation, Science and Economic Development Canada

Nanaimo bar (198) NoirChocolate/Shutterstock.com

poutine (199) Foodio/Shutterstock.com

Wonderbra (200) © McCord Museum, Montreal

flag colour standards (201) (fabric patches) komkrit Preechachanwate/Shutterstock.com

Sphynx cat (202) Eric Isselee/Shutterstock.com

Yukon gold (202) photogal/Shutterstock.com

Bovril (204) Chris Leachman/Alamy

derivative valuation (204) Bloomberg/Getty Images

multi-touch screen (205) OmniArt/Shutterstock.com

Juste Pour Rire (206) The Canadian Press/Denis Beaumont

internet search engine (206) doomu/Shutterstock.com

INDEX

barometer, Mount Logan, 54
Barthomeuf, Christian, 171
basketball, 184
Bawa, Riyadh, 202
Bear, Clayton, 31
Bechard, Jerome, 165
Beers, William George, 175
Belcourt, Christi, 89
Bell, Alexander Graham, 16, 48, 52, 53
Belleau, Bernard, 137
Bellini, Francesco, 137
Bell Laboratories, 165
Best, Charles, 121
Bethune, Norman, 122, 124
Bigelow, Wilfred, 129
Black, Fischer, 204
BlackBerry, 31, 67
Black Brant, 28
Blackburn, Elizabeth, 135
Black-Scholes formula, 204
Blake, Hector ("Toe"), 110
blind, aids for, 77
blood transfusion service, 124
Bloody Caesar, 164
blue box recycling, 80
boating
    disappearing propeller, 99
    Laser dinghy, 167
Bombardier, Armand, 58, 61
bomb sniffer, 112
Bovril, 204
bowling, five pin, 186
Boyd, Winnett, 60
Boyle, Robert William, 101
Boyle, Willard, 165
The Brady Bunch (TV series), 164
Braille, computerized, 77
brain surgery, 123
Bray, Tim, 30
Breakey, Norman, 74
Breakfast for Learning, 82
Breslin, Mark, 169
bridges, no repair, 117
Bristol Aerospace, 28
Brockhouse, Bertram, 25
Brooks, Harriet, 17
Brown, Alan, 122
Brownie camera, 149
Bruce, Ian, 167
Bruce, Sandra, 34
Brunton, David, 99
Brunton compass, 99
Buckley, William, 120

Buckley's cough syrup, 120
buffalo coat, 174
building supplies
    long-lasting concrete, 117
    Plexiglas, 157
    thermal windows, 188
Burtnyk, Nestir, 166
butter tart, 177

C
caisse populaire, 153
Callow, Walter, 74
cameras. See photography
Campbell, Thomas, 98
Canada, as union of colonies, 96
Canada Dry, 151
Canadarm, 64, 65
Canadian Armament and Development Establishment, 28
Canadian Car and Foundry, 23
Canadian Corset Company, 200
Canadian Dictionary of Biography, 199
Canadian Geological Survey, 11
Canadian National Carbon Company, 163
Canadian Observatory on Homelessness, 81
Canadian Pacific Railway, 50, 150, 185
cancer. See medical science
CANDU reactor, 133
canoes, 40–41
canola, 135
cardiac disease catalogue, 124
Carmichael, Robert, 170
Carroll, Tom, 158
catalytic RNA, 136
cats, Sphynx, 202
cattle plague vaccine, 141
caulking gun, 16
CBC Television, 161
Cech, Thomas, 136
celebrations: potlach, 70
Central Experimental Farm, 15, 148
cervical cancer test, 115
Chalk River Nuclear Laboratory, 25
Chalmers, William, 157
Champlain, Samuel de, 40, 176
Chaplin, Charlie, 153
Chapman, Christopher, 164
Chapman, Don, 142
Chell, Walter, 164
chemistry
    catalytic RNA, 136
    chemical bridge, 130
    DNA-based, 136

INGENIOUS

Edited by	Doug Pepper
Managing Editor	Kimberlee Hesas
Design	CS Richardson
Typesetting	Erin Cooper
Production	Carla Kean
Printer	Friesens, Altona, Manitoba

Typeset in Swift Neue, Brandon Grotesque, and Neutraface